LONDON

HISTORY OF THE URBAN ENVIRONMENT

Martin V. Melosi and Joel A. Tarr, Editors

LONDON

Water and the Making of the Modern City

JOHN BROICH

UNIVERSITY OF PITTSBURGH PRESS

Published by the University of Pittsburgh Press, Pittsburgh, Pa., 15260

Copyright © 2013, University of Pittsburgh Press

Manufactured in the United States of America

Printed on acid-free paper

10 9 8 7 6 5 4 3 2 1

Library of Congress Cataloging-in-Publication Data

Broich, John.

London : water and the making of the modern city / John Broich.

pages cm. — (History of the urban environment)

Includes bibliographical references and index.

ISBN 978-0-8229-4427-0 (hardcover : alk. paper)

1. Water resources development—England—London—History. 2. Water-supply—England—London—History. I. Title.

HD1697.G74L663 2013

363.6'109421—dc23   2012047752

Contents

Acknowledgments

I thank Amy Broich.

I also thank Professors Bill Baker, Laurel Carrington, Jan Dizard, Paula Findlen, Brad Gregory, Dick Judd, Michel LeGall, Martha McNamara, Paul Seaver, Bill TeBrake, and Eric Weitz. I also owe thanks to colleagues Matthew Booker, Dan Cohen, Jared Farmer, Jules Gehrke, Rick Lopez, Chris Otter, Carol Pal, Jonathan Sadowsky, Priya Satia, Lisa Sedrez, Renee Sentilles, Peter Shulman, Rachel St. John, and Rod Wilson. And I thank the Case Western Reserve University Department of History for many reasons. And I owe thanks to Rhea Cabin, Margaret Harris, Don Johnson, Marissa Ross, Kalli Vimr, and Emily Sparks. I owe great debts to Bill Cronon, John Hassan, Chris Hamlin, Martin Melosi, and Dale Porter that go underacknowledged in the body of this book.

I thank Stanford University and the Department of History there for all kinds of support, including research travel awards. I thank Greg Call and Amherst College for the same. And I thank Case Western Reserve University, including the W. P. Jones Fund, for research travel support. Thanks are also due the Ursu Family Foundation. And thank you to the University of Pittsburgh Press, including David Baumann, Cynthia Miller, and Ann Walston, as well as freelance editor Maureen Creamer Bemko.

This book benefited greatly from the Sanitary Reform of London Collection in the special collections department of the Stanford University library. These materials contain the Metropolitan Board of Works and Metropolitan Water Board waterworks libraries. As such, the collection includes both Joseph Bazalgette's working library and the collections of other agencies and corporations replaced by the Metropolitan Water Board. It is an astounding and still underused resource, and the staff of the Stanford library special collections department are an astounding group, too. Renowned archivist Margaret Kimball, now retired, was of critical help. I also thank the staff of the British Library at St. Pancras and Colindale, Historical Manuscripts Commission, London School of Economics Library, Surrey History Centre, National Archives (or Public Record Office, as it was still known when I did much of my work there), New York Public Library, and Wellcome Trust Archives. Special thanks are due the London Metropolitan Archives, where I performed the majority of the research for this book.

Peter Stansky, a student of David Owen, a historian of the Metropolitan Board of Works, began his long career as a scholar of Victorian political culture. It seems fitting that I was one of Professor Stansky's last PhD students. He is as gracious as he is expert, and I continue to learn from him.

It might surprise some that Richard White was the other main forebear of this book. I humbly aspire to adequately reward some day his patient teaching.

Introduction

In 1844, the engineer James Smith explored the alleys and courts of Leeds, stepping through and around the cast-off filth of the poorest of the city's 170,000 inhabitants. He encountered heaps of waste that lingered for six months and the stench of drains that lacked any flushing water for unknown spans of time.[1] Meanwhile, Dr. William Kay scouted Bristol, where only 5,000 of its 130,000 inhabitants enjoyed piped water, the remainder walking long distances to draw from public wells or, more often, simply going without.[2] James Martin investigated Leicester, where the sick suffered from a lack of water due to the scarcity of common pumps in working-class neighborhoods.[3] Smith, Kay, and Martin were members of the Commission for Inquiring into the State of Large Towns and Populous Districts, an official government investigation of the severity and causes of the health and sanitation problems more and more frequently seen, or smelled, in Britain's cities; its thirteen commissioners visited Britain's fifty largest cities and towns, met with local doctors and public health officials, and conducted a survey of each locale's water supply, water drainage and waste handling, working-class housing, and other living conditions.

They argued that Britain's cities, their populations having grown extraordinarily in recent decades, had not expanded their water and sewer capacity proportionately; water sources that were sufficient for the populations of previous centuries were stretched to their limits and threatened with the refuse of larger populations. There were few sewers as we know them. Water removal provisions usually aimed to shunt storm water away from structures and did not always include any sort of flushing except by rain. Private companies, which operated most cities' water supplies, tended to serve wealthier neighborhoods while bypassing the poorest. In the absence of effective central legislation or regulating authorities, these companies could not be forced to provide a constant service, nor were they all obligated to maintain a minimum standard of quality. The result, according to the commissioners' work in 1844–45, was a precarious situation that they believed threatened the health of millions; epidemic lurked in the poorly disposed waste of the masses, in the water supply tainted with it, and in a supply that threatened to fail at any moment, perhaps in the middle of an outbreak of disease. For the commissioners, it was rather straightforward to identify the intolerable state of affairs: insufficient clean water was being introduced into cities and insufficient polluted water was being extracted from them. In the eyes of investigators, a myriad of other urban problems would be solved if only this hydraulic input/output problem were solved. They contended that working-class dwellings would be cleaner, pure water would replace alcohol as a beverage, and the workers would wash more often, thus inhibiting illness.

Correcting the situation was less straightforward. Restructuring cities' water systems demanded urban governments with clearly defined and broad powers, and these were very rare before the turn of the twentieth century. But the 1844–45 commission and investigations like it, coupled with cholera outbreaks in 1848 and 1853, elicited impassioned newspaper columns, public debate, and, ultimately, environmental action undertaken by urban governments and endorsed by the national government. There was, in short, a transformation in government machinery in order to repair cities' hydraulic machinery. Cities sought and received the authority and means to purchase private water companies, and they borrowed large sums of money to construct sewers and build new waterworks. Action was widespread and profound even in an age that valued economizing. Between 1841 and 1881, the proportion of municipalities that took responsibility for providing their own water supply doubled, with more than 150 towns and cities adopting municipal water supplies.[4] In the same period, increases in the average water consumption per head ranged from around 60 percent to as much as 400 percent.[5]

The changes in Britain's urban governments were profound not only for the sheer magnitude of the water system reformation across towns and cities but also because of a new ideal that was prominent in the debate about identifying and solving urban water problems. Many of those who transformed urban water regimes acted on principle as well as pragmatism, basing their action on a vision of what they considered an enlightened—a modern—society and city. From the 1840s onward, a belief developed among urban leaders that a modern society would be one that equipped its cities with a physical infrastructure that ameliorated conditions dangerous to public health. Local governments had to act in the public interest, even if it meant significant expense to taxpayers. The editors and publishers of newspapers and periodicals often lent their support, registering indignation at the apparent water and drainage crisis, even calling the failures a "flagrant social crime" committed by a society that claimed nearly miraculous scientific knowledge and mechanical skill.[6] The nation, by tolerating the suffering in the deserts of its cities, was failing that test of "the progress of true civilization."[7] "Water reform," as contemporaries called it, was a moral obligation and modern prerequisite.[8]

From Belfast to Birmingham to Bristol there arose a consensus that cities should purchase the joint-stock water companies, which had failed citizens in burgeoning communities, and make water supply a matter of local government responsibility. A second but no less important component of this movement involved the construction of new waterworks. Water sources of greater volume and purity than existing ones were required in order to solve the problems of insufficient, impure, and irregular water service and the lack of drainage and sewers. The projects undertaken across Britain were quite similar both for practical reasons and because projects were based on shared ideals of modernization. The new model waterworks tended to be quite large, with reservoirs ranging in size from a few acres to the equivalent of Britain's largest lakes, so that, by 1880, various commentators described a "fashion for huge schemes." Such magnitude was due, in part, to local governments seeking to ensure the growth potential of their cities with one monumental project. The larger the water system, the larger their populations and water-consuming industries could grow. The same small group of engineers was behind the vast majority of the projects, too, contributing to the consensus about their design and scale. A few individuals became identified as the experts in the field; they were men whose names became connected with the more monumental water schemes and whose names lent prestige to the town council that hired them as a designer or consultant.

This book grew out of a simple question: Why, when every other large town and city in Britain took over the operation of its water supply in the nineteenth century, did London not do so?[9] The governments of Bradford, Birmingham, Leeds, Liverpool, Manchester—every provincial seat, in short—took on the responsibility of supplying its citizens with water in this period, buying local commercial suppliers or building new waterworks, but not the government of London. That one city should defy an overwhelming trend is noteworthy, and that the city left out of the development should be the first city of the nation, usually the epicenter of trends and movements, is remarkable. The search for an explanation for this paradox very quickly leads to yet more questions, more distant histories, and the need to reevaluate the implications of the primary question and even of water itself.

This line of inquiry revealed that the water systems constructed by British towns, or purchased from existing water companies and expanded, represented environmental reforms with broad implications. First, the development of water systems involved a significant development in the system of modern urban governance. Additionally, waterworks were viewed as not only a means of literally engineering public health but also as an essential mechanism for realizing a new epoch for the British city. Through water, the city could be modernized and moralized. This motivation belies the idea that the development of urban water systems was a simple, automatic process in which growing, industrializing towns of the nineteenth century faced epidemic disease and a lack of clean water and that the municipal waterworks that proliferated, first in Britain and then throughout Europe and America, were the inevitable, obvious response.[10]

This book, like a collection of histories written in recent years, is predicated on the belief that the story is far more complicated than it appears.[11] Figures ranging from legislators to self-proclaimed experts to water consumers and others argued over water reform at every step. These groups did not agree on the problems presented by insufficient, unclean water. They argued over critical questions: What was the relationship between disease and poor water supply? Were problems chiefly caused by dirty environments? These interest groups also failed to agree on the solutions. What were the appropriate technological responses? What expenses, what changes to landscape were justified? As this book insists, those who helped to hash out water reform were guided by their views on the proper order of society and morality, by their visions of the city and government, and by their conceptions of the proper relationship between people and the landscape.

What becomes clear is that tremendous social import was invested in the basic resources that sustained a city's life moment to moment and that it was usually during moments of environmental stress that this critical investment came into sharpest focus. At those times especially, when natural systems ceased to serve the social systems that relied upon them, groups within society sought to realize their goals for the community through the physical reorganization of the environment—of resource collection and delivery—and the administration of the environment. The environment drove political action, and society reengineered the environment to effect social change.

Society was not unified in its goals for the environment and itself. In the case of London, rival political authorities argued over the right to re-administer the city's environment—to operate the acquisition and delivery of water—in order to realize very different visions. One authority offered one picture of the modern city, and a rival power offered an alternative modernity. So, finally, the question that originally prompted this line of inquiry—why London's government failed to provide its own water supply—needs to be revised. How did water come to be so charged with meaning that it resulted in a bitter struggle in London, with unexpected consequences?

LONDON

CHAPTER I

Water and the Making of the Modern British City

THE GENERAL PICTURE of water supply in Britain before 1800 is varied, consisting of individual private sources, community sources, and small-scale commercial sources operated by single entrepreneurs or small groups more or less satisfying the needs of Britain's towns. Prior to the explosive urban population growth rates of the nineteenth century, there were far fewer difficulties in ensuring that sufficient clean water entered such towns. Communities could rely on drinking water sources that had been used for centuries: a local river or pond, a house well, and so on. Town corporations and charities usually organized modest shared supplies as well, in the forms of neighborhood wells and cisterns supplied by conduits from hinterland streams or springs, for example. The City of London built a number of conduits leading from countryside springs and streams to public cisterns from the thirteenth century onward.[1] Charitable patrons offered a fountain in sixteenth-century Manchester.[2] Glasgow's magistrates provided a scattering of wells in the eighteenth century.[3] There were also some enterprising individuals who saw an opportunity to provide a convenient service for profit in early modern towns. One of the oldest and most common forms of water enterprise was the simple water cart. A hardy individual would push or pull a cart holding a large barrel of water drawn from a reliable water source and sell to consumers who could not rely on a well or did not want to trek to it. More complex services involved conveying water from outside towns' boundaries and even piping water directly to consumers' buildings. For example, in the mid-seventeenth century, two men undertook to pipe water into Belfast from its nearby hinterland via hollow tree trunks.[4] One William Yarnold bought a license from Newcastle's town council to bore into a local spring in 1697.[5] In 1581, the City of London granted one entrepreneur a lease to one of London Bridge's arches in which he built a waterwheel that pumped Thames water (though it already had a less-than-spotless reputation) into a rooftop holding tank; it then passed to customers' premises under pressure in wooden pipes.[6]

Just as water supply arrangements were haphazard, so was the regulation of water suppliers and supplies. What little regulation existed was decentralized. Some town (corporation) charters included the obligation and right to protect local water supplies from misuse and pollution, for example; in other areas magistrates oversaw water sources by ancient custom.[7] However, it was extremely rare that local authorities had a legal obligation to supply water for drinking or sewerage or had the statutory means to enforce their rights.[8] Townspeople, then, could not expect a reliably sufficient or pure supply of water.

Reliability and purity were threatened by the demographic changes of the first half of the nineteenth century. Britain's population rapidly expanded in the period, and the numbers of city dwellers grew almost as quickly. The population of England and Wales doubled between 1800 and 1850, with Scotland and Ireland following close behind.[9] London's population nearly doubled between 1800 and 1840, while Leicester's and Manchester's populations tripled, rising from 16,953 to 50,806 and 96,000 to 313,000, respectively.[10] In 1800, no city besides London could claim 100,000 inhabitants; by 1840 there were five such cities in England.[11] By the middle of the period, a larger proportion of Britain's population lived in towns than in any other Western country.[12] Explanations for this increase in population are hotly debated, but among the principal contributors were a high rate of marital fertility in the growing cities and a decreasing death rate abetted by dietary improvements.[13] For the first time, too, towns ceased to be places where the death rate outstripped the birth rate; migration to cities, then, added numbers on top of the natural increase.[14] The industrial changes of the early nineteenth century invited such migration. There had been cottage industry, with families producing handicrafts to supplement agricultural work, since time immemorial, but more and more often in the late eighteenth century, entrepreneurs, especially in the textile industry, saw an opportunity to increase productivity by concentrating more and more labor under one roof, more closely supervising workers, and increasing output by locating machines at the workshops, or "manufactories."[15] The multiplication of such arrangements and the concentration of such operations in London and especially in northern centers played an important role in the Industrial Revolution.

The minimal arrangements that had more or less sufficed to provide town dwellers with water and deal with their waste began to fail under the pressure of urban growth and industrial demand on the eve of Victoria's accession to the throne. As Glasgow's numbers, for example, approached eighty-four thousand at the turn of the nineteenth century, the inhabitants could more and more often be seen standing in long lines to draw their daily water from communal pumps; more and more often that water was tainted with sewage from overcharged cisterns encroaching on the wells.[16] Even if traditional local sources had been able to supply expanding urban populations, they came under increasing threat from multiplying industries such as tanneries, dyeworks, and bleachworks.[17]

Something had to be done, but the vast majority of local governments were either unwilling or unable to take action. In some instances, town authorities simply had no legal authority to supply water; in many cases those rights were ill defined or limited.[18] There was certainly no general legal act delineating towns' rights and responsibilities in the matter. Constructing new waterworks sufficient to satisfy the demands of expanding populations and increasing industry would require large initial outlays of capital, as well, and towns simply lacked the power to borrow sufficient funds for sufficient terms during the period.[19] In addition, the prospect of increasing rates to service such loans was not an appealing one to town councils; besides a general desire for economy, especially in an age of growing economic liberalism, large property holders—the urban elite—foresaw an expansion of their taxes in proportion to the value of their properties.

Scores of local governments turned to the same solution in the first few decades of the century: they placed responsibility for water provision in the hands of joint-stock companies. Private companies did not have to contend with public reluctance if they wanted to raise rates and increase debt; they went directly to Parliament for power to issue stock, build works, and draw water from local sources. In Glasgow, a group of patricians formed a company for pumping water from the river Clyde, receiving parliamentary approval to raise £100,000 in shares of £50 in 1806.[20] Manchester, Leicester, Huddersfield, Newcastle—more than eighty towns in total—followed suit before 1850.[21] By 1828, London alone was served by nine such companies that parceled out service areas in the metropolis among themselves.[22] Only 10 out of 190 local municipal authorities operated their own waterworks by midcentury.[23] By giving companies the right to operate within town limits, municipal authorities hoped that they had averted an impending crisis that they were powerless to prevent on their own.

The Bases for the Growing Power of Towns

Prior to 1835, there was no such body as a representative town council in England or Wales. Towns were administered instead by a variety of individuals and organizations. First, there was the local justice of the peace or JP, appointed by the Crown (having been nominated by the country's existing justices of the peace) to preside over the county and deal with, in addition to judicial matters, licensing, game law, building and repairing roads and bridges, and collecting local taxes. Then there was an ecclesiastical sort of area administration. The parish was the smallest administrative unit of the Church of England and the sole responsibility of a rector, vicar, or curate. Within each parish met the vestry, a council of men who took care of certain parish duties, including the administration of the relief of the poor, but the powers and abilities of the county's justices of the peace and its vestries were no match for the special circumstances of growing towns.

The administrative needs of the 178 officially recognized towns were addressed in no single manner.[24] Townspeople usually, though not always, had received from the Crown a medieval charter that often made provision for the election of town officers who served on a corporate council. A large majority of these were not democratic but closed bodies, most often the reserved domain of the local Tory oligarchy.[25] Depending on the extent of their royal charter, these officers could impose taxes or rates and duties, pass bylaws, and exercise other rights. The charter also freed townspeople from the jurisdiction of the JPs, removing the town, in effect, from royal oversight. If the charter granted the town the status of a borough corporation, the town would have its member of Parliament or MP. Town corporations did not exist to represent every member of the urban community; they were instead legally recognized organizations invented for the mutual benefit of their members.[26] The corporations usually had responsibilities similar to those of the JPs, but street cleaning and drainage, dealing with the poor, and crime prevention took on more importance. Corporations, not bound by any general governmental act spelling out their duties, did not take on all responsibilities equally, so ad hoc groups often played many roles within the town. A body of drainage commissioners might deal with storm sewers; street lighting commissioners might look to lamps. On the other hand, commercial enterprises might provide a gas or water supply. Some towns never received an incorporating charter, in which case ad hoc bodies and commercial services supplied the only administration in the town. Parish vestries continued to fill their limited role in incorporated towns and within corporations.

In 1833, the Whigs appointed the Royal Commission on the Municipal Corporations of England and Wales to investigate how well the nation's growing towns were being administered. In fact, the Whigs already knew the answer: municipal corporations were not fulfilling their duties and were renowned for a tendency toward inertia at best and outright corruption at worst.[27] The Whigs also knew that the municipal corporations were closed off from the rest of their communities, strongholds of the landed interest and their political rivals. The Whigs saw cracking them open as a natural postscript to the Reform Act of 1832, which nearly doubled the size of the kingdom's electorate and realigned parliamentary constituencies to recognize the growth of the towns. The Whigs succeeded. The Municipal Corporations Act of 1835 eliminated all existing corporations—the City of London's local government excluded—and created town councils in their places. The triennial elections for the councils would be based on household suffrage, with the requirement that the head of household had resided and paid rates within the town for the previous three years. Councilors would elect a mayor from among their fellow members and aldermen—who would constitute one-quarter of the council's number.[28] While the ostensible goal of the Whigs was to crush the Tory local stronghold, they expected Tory oligarchies to be replaced by Whig-Liberal oligarchies.[29] They were not disappointed. Ordinary townspeople could not sit on the new councils, a substantial property requirement barring their participation, but a new class of individuals did overturn the corporations' status quo from the first council elections, with the upper-middle-class Whig-Liberal bourgeoisie seizing power.[30]

The Municipal Corporations Act had the practical effect of turning out the old power holders and providing a new standard constitution for towns that had previously been incorporated, but it did not spell out many new powers for the councils. It provided for policing authorities but otherwise offered only a blank slate. The act's Whig promoters, however, had high expectations for their new councils: these new bodies had provided a constitutional framework on which the councils could build. The act also provided a mechanism by which previously unincorporated towns could petition the Crown for a charter and be immediately included in the terms of the act—an opportunity of which communities rapidly took advantage.[31] One way councils expanded their range of responsibilities and power was through general parliamentary acts; for example, in the 1840s, Parliament passed the Towns Clauses Acts, which offered model clauses that councils could adopt.[32] The Public Health Act of 1848 enabled towns to operate as local boards of health with powers of enforcement. But much more often, town councils expanded their powers by applying independently for a local private act from Parliament.

The expansion of local government power through this chief mechanism had an environmental basis. The first steps councils took to acquire authority were for the purpose of correcting problems councilors saw with their own eyes. Upper-middle-class councilors walked down dark streets and sought authority to provide a gas supply; they saw filth and sordidness and sought power to operate street cleaning, supervise lodging houses, and provide public restrooms. And, almost universally, they worried that the towns' slums incubated disease, so they pushed for power to operate a water supply and provide sanitation. The majority of the first bills that councils submitted to Parliament were for sanitation and often called "improvement" bills.[33]

Describing a Perceived Environmental Crisis

In the late 1830s, an austere civil servant named Edwin Chadwick picked his way through the muddles of Britain's new industrial cities investigating the circumstances under which the nation's average city dwellers lived.[34] He was appalled at what he saw as the deterioration of the health of Britain's large working class, and he sought ways to improve the situation. He had absorbed his mentor Jeremy Bentham's vision of the Utilitarian state, a vision of an active, efficient, and empowered government that took bold measures to effect the greatest improvement for the greatest possible number of citizens. On Bentham's advice, Chadwick became a dogged collector of data and narratives, the very model of the Victorian social investigator. Chadwick became a sort of Victorian Argus, looking for blight in all directions through an army of deputized informants as he prepared to promote the sort of bold social fixes that Bentham had envisioned.[35]

Chadwick would later become one of the most influential administrators in British history, but, in the late 1830s, he was simply an active but unknown public servant conducting an investigation for the Poor Law Commission.[36] The investigation was semiofficial, unfunded, and, until he presented his report, disregarded by his superiors. His conclusions defined the terms of social investigation and urban transformation in his generation.[37] What Chadwick discovered pointed to one particular factor as central to the health problems of the nation's common citizens: the lack of sufficient water meant that working-class dwellings and bodies went unwashed; it meant "the constant retention of pollution of several hundred thousand accumulating in the most densely peopled districts," and the absence of uncontaminated water meant working-class inhabitants would continue to rely on alcoholic beverages to slake their thirst.[38] The water shortage appeared as unexpected as it was pervasive. "No previous investigations," Chadwick wrote, "had led me to conceive the great extent to which the labouring classes are subjected to privations . . . of water for the purpose of ablution, house cleaning, and sewerage, . . . drinking, and culinary purposes."[39] In his final report, Chadwick identified other problems of the overcrowded, poorly ordered city, from airless, cramped working-class housing to the dangers of locating cemeteries in proximity to housing. The greatest positive effect, he argued, could be realized by addressing the water dilemma.

For Chadwick, the problem of public health was entirely an environmental problem. He and many of his contemporaries identified proximity to filth, unwholesome atmospheres, and miasmas as the cause of disease. Chadwick wrote that "there is no one point on which medical men are so clearly agreed, as on the connexion of exposure of persons to the miasma from sewers, and of fever as a consequence."[40] According to miasma theory, waste was supposedly rendered innocuous by submersion under water; building drains and sewers, in short, would serve to protect the public health far more than would a legion of doctors. "The great preventives—drainage, street and house cleansing by means of supplies of water and improved sewerage," he wrote, "are operations for which aid must be sought from the science of the civil engineer, not from the physicians." Flush the environment, argued Chadwick; get clean water into and polluted water out of the city.[41]

Chadwick presented his evidence and argument for environmental management in a substantial report under his own name in 1842, and though his was only a semiofficial inquiry, his findings won significant public attention.[42] His report sold eight times as many copies as any other official publication of its day, it led to the formation of the Health of Towns Association, a humanitarian activist group, and it forced the Conservative Party–controlled government to call for a full inquiry into the question of the state of cities.[43] Chadwick lent this environmental approach to social problems to the Royal Commission on the Health of Towns; it too looked particularly to hydraulic factors as the main determinant of health and mortality rate. The thirteen commissioners included the civil engineers Thomas Hawksley and James Smith and the railway engineer Robert Stephenson; their favored interviewees in this royal commission report, too, tended to be other engineers and builders.

An Indictment of Water for Profit

The thirteen commissioners, reporting in 1844 and 1845, pointed a finger at commercial water suppliers for contributing to the urban environmental crisis. First, they charged, joint-stock companies had not secured the new water sources demanded by expanding urban populations. The companies' main water sources were often the same communal sources that had served towns for generations but that began to fail when they could not satisfy the needs of industrialization.[44] Liverpool's water company, for example, provided only a "stinted and intermittent supply," the commissioners reported, which resulted in "sparing use, so that a proper degree of cleanliness in the houses of the poor is prevented."[45] They reported that in towns with commercial water suppliers, water tended to be "very deficient," often with water supply systems extending only to wealthier areas and "the poorest and most populous portions deriving little or no benefit." They offered the example of Birmingham, with only 20 percent of houses receiving water, and Newcastle, with only 8 percent served.[46]

Simple principles were behind this state of affairs. Extending water service throughout cities called for additional outlay of capital to lay pipes, increase steam pumping, and so on; any outlay of capital could lower shareholder dividends unless the capital was balanced by the return on the investment. Water was not sold by the gallon, so it seems the companies prospered most merely by increases in water rates stemming from the steady rise of the rateable value of the houses served.[47] Companies were more interested in providing their product to reliable, more affluent customers—living in finer houses—instead of expanding supplies and services to larger populations. "The Directors of such Companies naturally hesitate to carry pipes into districts, where the returns for the money expended are so precarious," concluded the commissioners, and those directors "seldom consent to supply the houses of the poor unless the landlords become responsible for the payment of the water rates."[48] Landlords, in turn, were unlikely to assume responsibility for rates since doing so would require raising rents and since itinerant industrial workers, who tended to rent by the week, could depart overnight.

For the same reasons of simple economics, profit-seeking companies frequently provided water of dubious quality. In the hands of commercial suppliers, water became the most elementary of commodities. Companies did not make more profit per unit of clean water supplied than per unit of dirty water supplied. The commissioners found frequent instances of pollution—from Newcastle, where the company's waterworks were situated near the outfall of a common sewer, to York, where taps ran muddy after heavy rains.[49] Since most towns were served by one monopolistic company, water suppliers could "dispose of it only to those persons who are willing to buy it at such rates, and on such conditions, as they are pleased to impose."[50] And often companies imposed water of an unsavory condition on consumers.

Though the commissioners were hardly radical reformers—they were headed by a Conservative peer, the Duke of Buccleigh, Sir William Cubitt, a Conservative MP was also a member, and the remainder were sober professionals—the commission's conclusions were forceful and broad ranging.[51] That force derived from the simplicity of their conclusions: the commissioners blamed the problems of water supply on the harmful influence of basic economic principles, and thus their objections applied universally to commercial water operations. Where there was a profit motive, the seller necessarily sought to provide as little as possible for the best possible return; add stockholders to the equation, and there was another party whose interests came before those of consumers. The commissioners quoted one of their correspondents, who reached the same conclusion: "Whatever is made a matter of sale, presupposes two parties having opposite interests, viz., the seller and the buyer . . . [and] the buyer is necessarily at his mercy."[52] Because the problem was so fundamental and all-encompassing, concluded the commission, so too must be the solution. "A copious supply of pure water cannot be secured to the poorer classes of the community," they wrote, "unless the duty of providing it is placed under the management of the local administrative body . . . to ensure its regular distribution."[53] Private enterprise—and market forces—had failed to meet the challenge of urbanization and industrialization; local state authorities, instead, must safeguard the interests of their citizens.

From another perspective, the state, embodied by Chadwick and the royal commission, found its citizens unwilling or unable to safeguard *its* interests. Cholera struck the country in 1831 and typhus in 1837; the government's Registrar General showed alarming rates of mortality in Britain's cities.[54] To many observers, unwatered, unwashed cities threatened to unravel, or at least disastrously foul, the social fabric. The solution, in the words of the 1844–45 royal commission, was to "promote habits of cleanliness among the population," but the commissioners complained that authorities could not compel the working class to use more water unless it was "readily accessible at all times, without trouble."[55] Only with universal, affordable water service could city dwellers and landlords be required by law to connect their properties to water supplies. Only if the state better managed natural resources could it better manage the behavior of its people and, ultimately, preserve itself.

The Ascendance of an Ideal

The royal commission of 1844–45 advocated the takeover of commercial water supplies by local authorities, but it was unlikely that Parliament would respond by mandating municipalization across the kingdom. The prevailing economic and political philosophy regarded private enterprise as sacrosanct, and those who espoused such philosophies were wary of stringent regulation and centralization of government authority.[56] With another cholera epidemic bearing down on the country, though, Parliament did pass the Public Health Act of 1848. It called for local authorities to establish boards of health to monitor urban environmental conditions.[57] It also explicitly granted town councils the power to borrow money to make sanitary improvements and gave them the authority to purchase water companies—if those companies consented.[58]

While it did not result in sweeping legislation, the work of Chadwick and the royal commission was ultimately influential in the way it focused attention on the environmental problems of towns within prominent social circles in the towns themselves. The 1844–45 commission, with its legal authority to compel evidence from Britain's fifty largest populations—an authority of which they made good use—literally forced that urban elite to focus their attention on public health threats, even if those individuals were part of the minority who enjoyed piped water and adequate waste disposal. The commissioners themselves certainly believed in their influence. "By calling to our aid the assistance of the most influential and intelligent of the inhabitants," they reported, "[we] exposed to their view scenes of misery and neglect, of which many were previously ignorant[,] and directed their attention to causes of disease."[59] The aftershocks of the commission's reports rippled through the conduits of public discourse, as well. Abridged versions of the royal commission's investigations of individual cities and regions were sold in pamphlet form, public health agitators issued their own calls for action (declaring themselves vindicated by Chadwick and the commission), and newspapers repeated the lurid tales of urban squalor.[60]

Town councilors repeated the commissioners' criticism of the water supply status quo. Joint-stock companies had not solved Britain's urban water crisis; rather, they were fundamentally predisposed to offer deficient measures. For instance, in 1852, Bradford's borough councilors complained that the town's sole water company "had not developed all the benefits" of a water supply, instead preferring to pay shareholders a 9.5 percent dividend.[61] As they had for the royal commission, Bradford's objections to commercial water supplies came down to elementary economic principle. "When water was under control of private companies, the chief desire of the directors was to obtain good dividends," said one town councilor. "When the Town Council possessed the works," he explained, "instead of looking for revenue beyond what would maintain the works in efficiency, their chief object would be to make the works instrumental to the promotion of cleanliness, the health, and the comfort of all classes of the inhabitants."[62]

The same year, Glasgow's town council, as another example, came to the same conclusion as the royal commission regarding public control of water supplies; it unanimously resolved to operate the water supply "as a separate public trust for the benefit of the inhabitants," rejecting water supply via "any private company."[63] In 1851, the northeastern industrial town of Darlington also came to the same conclusion so many others had: the local water monopoly, as the town's sanitary engineer reported, charged too high a rate to compel inhabitants to accept a piped supply, so until the town itself provided water, authorities could not mandate universal water service and drainage.[64] The Manchester Borough Council reached this conclusion as well, resolving that it must "obtain power to accomplish the distribution of water at cost price to inhabitants" so that it could "carry into effect within the borough the recommendations of the Sanitary Commissioners."[65]

The cholera bacterium intervened to heighten the sense of a water crisis as epidemics revisited Britain in 1831, 1848, and 1854. Dr. John Snow first linked the transmission of the disease to water supply during the 1848–49 epidemic, which, at its height in the late summer of 1849, ended approximately one thousand lives per week.[66] When the disease struck in 1854, Dr. Snow was again at work studying the spread of the illness, comparing the incidence of cholera among the populations served by two of London's many water suppliers, the Lambeth Company and the Southwark and Vauxhall Company. He found a far greater rate of infection and mortality among the Southwark and Vauxhall customers, and his conclusion offered another indictment of the commercial water supply system: the company that had willfully failed to invest in filtration and secure sources away from urban populations was directly responsible for the deaths of its customers.[67]

Chadwick and others had already deduced a link between cholera epidemics, water, and the spread of illness, not because they blamed germs—Dr. Snow himself did not blame germs—but because they knew that the waste of the ill could both poison water supplies and contribute to miasmatic conditions.[68] Whatever the exact connections, water supply was now even more a matter of immediate life and death as cholera bore down. Warily watching as the disease crept toward Britain from Europe in 1852, a newspaper editor declared, "Nothing but good sanitary arrangements can save us."[69] And on the eve of cholera's arrival in Britain, the Lord Provost of Glasgow declared that if "[the] community were fearful of that scourge, Asiatic cholera, again visiting this city, the great aim and object of them all should be to secure an abundant supply of pure water."[70]

Cholera was certainly one factor that motivated change, but cholera could never determine the outcome of the water reform movement. If cholera "forced" attention on unsatisfactory water supplies and drainage provisions, or if it "led to" municipalization and new waterworks, the precise ways it did so were diverse, changing, often indirect, and often ambiguous.[71] Cholera could not change society, create technologies and management systems, or reorder urban politics. Only human actors could do that. *Vibrio cholerae* was a microscopic, comma-shaped bacterium that meant nothing to humans if it did not make them sick. The life cycles of bacteria combined with technologies, the movement of water, the action of weather, human ideas about morality, various theories of contagion, and other factors to effect change in the city.[72]

Germ theory was not necessary, in any case, for middle-class observers to perceive a health threat from the lower classes. The middle and upper classes' defense against epidemic was to make certain that the working class was supplied with good water and adequate drainage. In 1847, the editors of the *Ipswich Journal,* like many others, took it as a truism that it was in the working-class home that "pestilence first finds its footing, and from which it issues to produce disease and bereavement in the homes of the higher classes." The writer(s) took the situation as providential, since it forced the rich to concern themselves with the sanitary state of the poor. With "the cholera . . . on its destructive march towards our doors," the rich had better make certain "the humblest abode was . . . supplied with pure water."[73]

Within a decade or two, a consensus spread among social commentators and local officials as objections to commercial water suppliers and advocacy of public control as a better alternative became a standard refrain. The Tory mayor of Leicester, for example, decried weighing "the lives and health of Leicester" against "the question of income."[74] Thomas Avery, a Birmingham town councilor, agreed that "it is the duty and object of the ["trading organizations"] to protect the interest of its [*sic*] shareholders; but of the ["Municipal Government"] to consider the general welfare of all classes of population."[75] W. T. Gairdner, a professor of medicine, declared in a lecture in 1862 that "the moment water becomes a *commodity*—an article having commercial *value* in the ordinary sense of the term—you have reasonable grounds for suspecting that the community has culpably and negligently abandoned its rights, to the great danger . . . of the public interest."[76] Thus, communities should provide themselves with water—never a private company.[77]

By 1881, the economist Arthur Silverstone could describe a "movement in favour of the acquisition of water works by local authorities."[78] First Manchester, then Leeds, then Glasgow, then Birmingham municipalized their waterworks, with the movement reaching its peak between the mid-1860s and the end of the century, when approximately 40 commercial water enterprises were taken over by town authorities and many more were newly instituted.[79] In total, approximately 180 towns either purchased or built their own waterworks between 1845 and the year of Silverstone's survey, and the trend continued.[80] The percentage of towns supplied by municipal waterworks had thus risen from 41 to 80 percent in a forty-year period.[81]

Still, this was not a universal movement. In smaller, more easily supplied communities, private companies sometimes did succeed, and some new ones were established even in the prevailing climate of municipalization. In 1910, there were still 284 private companies compared to 821 public water supplies in Britain, though the remaining private suppliers usually served minor towns.[82] Even in 1973, when 10 regional water authorities came into being, there were 28 private companies holding out.[83] In each city or town that considered municipalization and building new waterworks, the debate over the wisdom of the move was renewed. Of course, those debates about local government expansion were most elemental among pioneers in the movement, but they never ceased being rehearsed. Municipalization was never automatic or taken for granted. In the late 1880s, the merits of granting Sheffield the right to offer a municipal water supply were hotly debated in Parliament.[84] That effort failed before ultimately succeeding.[85] Victorian urban society was no more politically homogenous than our own. The case of London's debate over municipalization will prove that beyond any doubt.

Local government expansion was in part rooted in particular problems arising out of the industrial environment. There may not have been a theory of collectivism in 1840 to rival that of laissez faire, but none was required; a powerful movement arose in part out of practical responses to problems associated with laissez-faire policies and the primacy of free enterprise.[86] This somewhat paradoxical development helps explain why the challenges of the urban industrial environment, water issues chief among them, were not addressed by regulating water companies. Certainly, the prevailing economic and political orthodoxy placed a high value on private enterprise—on undertakings like water companies—and so the government might have protected them from public takeover by regulating their performance.[87] But under what agency and with what enforcement powers would they propose to do so? And even if there had been an effective institutional framework available, which there was not, the political climate was hardly inviting of novel central control.[88] From the point of view of local communities, regulation from the center would have represented pernicious meddling. Town councils wished to address local concerns with local action based on local environmental knowledge. Popular feeling raged against Chadwick's proposal for a strong central board of health in the very same years that town councils were seeking strong powers of local sanitary control.[89]

The Modern-Moral Government

Another critical reason that communities did not simply seek better water provision through regulation was the ideological basis for their water reform efforts. For many, the endeavor to municipalize water supplies was motivated by more than companies' poor service; from the first years of the municipalization movement, many advocates saw their efforts as part of a campaign to reform government and society by transforming the urban environment—as part of a campaign, even, to produce a higher phase of British culture. This reformed society would be guided by a "public morality," a more conscientious set of principles more energetically enacted by government. Supplying water for drink, drainage, and washing—or, to be more precise, demanding that citizens accept and use more water—would help realize the modern-moral city.

Water supply was at the heart of this vision. The editors of the *Bradford Observer* explained that "it is not, therefore, consistent with public morals, any more than with public health, that either air or water should be dealt out in diminished quantity or inferior quality. . . . The bounteous Creator has given us of water, as of air, an inexhaustible and spontaneous supply, *it ought to be as free as air* [emphasis in original]; no expense, save that which is indispensable to bringing it within reach of the consumer, should be tolerated. [Poor water supplies] are a great sin against the public, and a standing memorial to bad statesmanship, and of an imperfect social economy."[90]

In Glasgow, too, reformers, called for a new "moral government" and equated the physical disfigurement of the city to a moral affliction.[91] In this same early period, the *Edinburgh Review* admonished its readers that they had no right to consider their society modern except as far as it took moral responsibility to maintain an unpolluted urban environment. Ensuring that cities were drained of wastewater demanded the rejection of the inactive governmental approach of earlier ages: "The progress of true civilization indeed, is best marked . . . by the facility with which men may crowd together into large city communities, without suffering from . . . the pollution which such accumulations would naturally create." It continued, "Like most other great benefits, it will be better accomplished through enlightened and well-weighed legislation, suggested by skillful minds devoted to the task, than by the blind chance which has hitherto ruled it."[92] With the municipalization movement well under way in 1869, a Birmingham town councilor directly linked the modern public-moral ideal and the expansion of state authority—and of environmental control—required to realize that ideal. "Surely it is the duty of a wise local government to endeavour to surround the humbler classes of the population with its benevolent and protecting care," he wrote in an address to the council, adding, "The compulsory supply of water to the poor appears to be the only effectual remedy for the evils now under consideration."[93]

It was an era in which middle-class legislators and experts sometimes portrayed their activism in a light similar to that favored by colonial "improvers." Henry Mayhew famously described what he called "our own heathen": the "creedless, mindless, and principleless" denizens of darkest London. Why venture to the other side of the globe to civilize savages when London's laborers had sunk to the "lowest depths of barbarism?," he asked.[94] George Sims wrote that, "in the matter of water and air, the most degraded savage British philanthropy has yet adopted as a pet is better off than the London labourer and his family."[95] And some of the same solutions for uplifting the foreign "savage" or "native" could work for the "moral elevation" of the native of Britain.[96] Education, for example, could "civilize or indeed humanise the labouring classes" of remote English counties; railways, too, would civilize people and "open up the country."[97] New water technologies could do the same. The urban poor were "wretched beings" brought low by the "physical debasement of their abode," "native and to the manner born," and enervated "by the malarious languor of the place," wrote the editors of the *Observer*. But cleanliness is next to godliness, they went on, and a constant supply of clean water was key to "the civilisation of the poorer classes."[98] In 1844, Dr. Thomas Smith, a commissioner of sanitation, made a similar case for water's power to have a "humanizing influence" on those in "the lowest state of civilization." Through water and cleanliness, they could be lifted up from savagery and brutality and made sober and peaceable.[99] In the colonies, engineers and officials would make the same case for the civilizing power of water systems, but there, they said, water supplies could correct the shortcoming of being the wrong race rather than the wrong class.[100]

The construction of new works frequently followed a city's municipalization of its water supply; a main objective, after all, of replacing the companies was to secure a larger, cleaner water source than those companies had provided or utilized.[101] Of the towns that municipalized water supplies between 1845 and approximately 1880, a majority developed new water sources and works immediately or within a few years of purchasing existing water companies. The acts authorizing municipalities to purchase the companies' undertakings usually included power to buy land, secure rights to new water sources, and fund new works.[102] In addition to permitting governments to demand compulsory water connections, developing larger supplies gave cities another opportunity to demonstrate their improved, modern character. "A good water supply is a necessary condition to high civilization," argued a correspondent of the *Medical Times and Gazette,* adding that a "modern society . . . also rates human life more highly."[103] In the words of the *Newcastle Chronicle*, the history of waterworks was "a record of the advancing prosperity and civilization of society. [We can] trace the progress of undertakings of this nature, and . . . contrast the simple operations which sufficed for the necessities of one age, with the bold and extensive projects demanded by . . . the present age of social progress."[104]

Through the technology of "bold and extensive" water projects, governments could ensure the future growth of populations and industry. They could, in a way, manufacture permanence and prosperity for their communities. "The health of our populations and the growth of our commerce are due, in great measure, to the abundance and purity of its supply," pled one engineer, Dale Thomas, to the Kingston-upon-Hull Borough Council, then considering abandoning its traditional well-drawn supplies. "Great and lasting benefits," he stated, would accrue only if "the source of supply is of a certainty that can be depended on both as to quality and quantity, [sufficient for] the most distant posterity."[105] A Bradford alderman warned that "on a sufficient supply of water depended in a great measure the commercial prosperity of the town. If they were going to maintain their position, they must take care that they had a larger supply of water than they hitherto had."[106] Such sentiments were echoed from Birmingham to Glasgow, where town officials called for water supplies of unprecedented "abundance, permanence, and regularity" and for new waterworks that would even give their manufactures greater value over competitors' facilities.[107]

Like competitiveness, pride was an important motivation for local governments to municipalize and construct new works. Securing permanence and plenty for future generations was an opportunity for individual officials and governing bodies to show their foresight and worth. In Huddersfield, for example, the mayor called on the town council to "avail themselves of [a new] supply . . . and accomplish a great thing—one worthy of the Town Council."[108] On the completion of Glasgow's new waterworks, a member of its water committee modestly declared the new system "one of the noblest municipal schemes ever, it is believed, devised and executed for supplying to a large and important commercial and manufacturing community one of the first necessaries of life." And the project would provide "future ages an enduring testimony to the enlightened wisdom, enterprise, and zeal of the Civic Rulers by whom they were . . . ultimately brought to successful completion."[109] Advocates of municipalization truly did praise governments that secured new water supplies, as the *Bradford Observer* did when the Bradford Town Council succeeded in buying its local company and arranged a major expansion of its waterworks. It promised that "future generations will gratefully acknowledge the benefit which this scheme has conferred upon the inhabitants."[110]

Other supporters of municipalization believed that new water projects would incite rivalry born of civic pride. "The ultimate good which may be looked for from such examples as Glasgow and Liverpool is incalculable," wrote Dr. W. T. Gairdner from Edinburgh, adding that "it is to be hoped that a spirit of generous rivalry will possess all our great cities."[111] Meanwhile, the *Edinburgh Review* in 1850 offered a list of towns, from Greenock to Paisley, that had municipalized and expanded their waterworks, as an admonishment to the slow-moving Londoners.[112] And, perhaps most rousing for the many Britons who viewed their nation as the advance guard of modernity, municipalization advocates pointed out that European and American cities were constructing impressive waterworks of their own. Sustaining its harangue, the *Edinburgh Review* editors wrote that

> it will be disgraceful to the nation if in our very metropolis we are surpassed in the arrangements for securing health and common decency, not only by the young republics of the New World, but even the ancient empires of the Old. We boast of our wealth, our freedom, our science, our powers of combined exertion, our sense of comfort, and our love of cleanliness; we glory in our civilization, but our glory becomes our shame, if still we are last in the race of humanity. The City of New York has expended 2,500,000 [dollars] on the Croton Water Works. . . . This is an exertion of which our Transatlantic brethren may well be proud.[113]

Indeed, advocates of new waterworks frequently pointed to New York's Croton scheme, completed in 1842, as a model project. It was large, far-sighted, beyond merely sufficient.[114] New, extensive waterworks, in sum, were conspicuous symbols that any town or city that built them was part of a new, advanced culture instead of the infamously careless one that had gone before. "Not in this country only, but in America and on the continent of Europe," lectured one engineer in 1856, "the waterworks of modern times are amongst the largest, the boldest and most successful productions of the age."[115]

Monuments to a Modern Age

Construct "large and bold" waterworks, and a city declared a bold policy. The larger and bolder the undertaking, the more pronounced the commitment to progress. Upon municipalization, cities tended to collect water from more sources and wider gathering grounds, to construct more impressive aqueducts, and to build larger filtering works. Towns that municipalized their water supplies between 1848 and 1880 invested more than £8 million in improving companies' existing works.[116] Upon municipalizing its water supply, Huddersfield, for example, immediately spent nearly £700,000 just on the expansion of its water gathering grounds.[117] Manchester constructed large new reservoirs and began importing the water through a ten-mile aqueduct in 1850.[118] In 1859, Glasgow began importing water thirty-seven miles from a partially artificial lake.[119] By 1880, an observer had identified a "fashion for huge . . . schemes."[120] A consensus developed among local governments regarding the fundamental design of new water supply systems, and a monumental scale was common to this method. By approximately 1880, then, most local governments desired a large water supply system of this single accepted method—with its conspicuous modernity.

This predominant method began with officials identifying and selecting a large water source in a town's hinterland, well away from settlements. This source might be a river, a lake, a group of streams, a number of springs, or a combination of these. Whatever the sources, the method required that the water be accumulated and reserved in large volumes. Streams might be diverted into an artificial reservoir, a lake might be embanked to hold additional water, or a valley through which a river passed might be dammed at one end. This supply of water was to be held at a significant height above the city to be served. If there were no suitable hills in proximity to the city, engineers simply searched farther and farther afield—even more than one hundred miles, if necessary—for the perfect elevated sources. The prevailing technique then called for an aqueduct or iron pipeline to deliver the new water supply to the city in need. Because the water flowed from a height, no pumping was required to convey the water over a distance. Owing to this feature, such schemes were often called "gravitation schemes." The aqueducts were often the impressively elevated kind that brought to mind Roman aqueducts, and the pipelines were occasionally driven through entire hills.

The gravitation system came to predominate in part because a coterie of influential engineers promoted it to town officials throughout Britain. For example, both James Simpson, vice president of the Institution of Civil Engineers, and Thomas Hawksley, a member of the Royal Commission on the Health of Towns and one of Britain's most celebrated hydraulic engineers, promoted a number of plans of this nature.[121] The engineer who promoted and constructed the vast majority of these schemes—as many as fifty—was John Frederic La Trobe-Bateman (1810–89). Though he did not develop the model, it became the quintessentially modern public work in his hands.[122] Bateman had learned the engineering trade as a surveyor assisting in the creation of millponds.[123] Manufacturers and canal companies demanded large volumes of water held in reserve, and those ponds, which had existed since the Middle Ages, needed to be expanded in scale as industrial activity expanded in the first half of the nineteenth century. In his early career in the 1830s, Bateman was called on to enlarge and build more of these reservoirs, projects that included raising the height of dams, calculating the capacities of watersheds, and building water conduits. It was a simple matter to adapt the arrangement to the provision of drinking water when a demand arose for it. Edwin Chadwick himself wrote Bateman in 1844, imploring him to become a specialist in the adaptation process.[124]

Bateman took advantage of the opportunity. From the second half of the 1840s through the next several decades he submitted several gravitation schemes per year to local authorities throughout Britain.[125] These were a far cry from the millpond arrangements he had designed as a young engineer. The scale of his municipal water projects was vast, the gallons of water per head he delivered dwarfed earlier supplies, and the cost of his projects, of course, increased proportionally.[126] He became well known as a master of the craft, and his was the name officials wanted connected with their towns and their projects.[127] His greatest projects, the Longendale and Thirlmere reservoirs for Manchester and his Loch Katrine system for Glasgow, led to the construction of some of the largest lakes in Britain, represented some of the most considerable public expenditures of any kind of their age, and represented a shift from waterworks as mere inconspicuous infrastructure into something monumental. Glasgow spent 1 million pounds to build a reservoir nine miles long and one mile wide at its widest.[128] Queen Victoria spoke at its dedication in 1859, praising the work, "which, in its conception and its execution, reflects so much credit upon its promoters, and is . . . worthy of the spirit of enterprise and philanthropy of Glasgow."[129]

The practical needs of exploding populations and forward-looking town officials invited such monumentalism. The growth of future generations simply demanded a larger volume of water than companies' existing sources provided. The Leeds Borough Council asked the engineer Edward Filliter to report on the best means of improving the town's water supply in 1866. He suggested securing no less than twenty million gallons of water per day, and, "for so large a quantity . . . no springs nor wells can be depended upon. . . . Rivers (or Lakes) receiving the rainfall from extensive watersheds must be resorted to, and if no watershed large enough . . . can be found, recourse must be had to storage of the floods of rainy seasons."[130] And in only a brief period, Leeds had constructed reservoirs within some of the watersheds Filliter recommended, while Bradford appropriated others nearby.[131]

Only a decade or so into the movement toward large gravitation projects, engineers began rejecting long-standing sources of water on the grounds of a general mind-set within the engineering culture. Certainly, many older sources were located in proximity to growing cities and thus were liable to more and more pollution, but Bateman and his fellow engineers offered a broad rejection of any source near a city. The old sources, they argued, were like the old mode of water supply—unsatisfactory at present and liable to experience disaster in the long term. Bateman, already a recognized authority on gravitation water systems, offered a general indictment of lowland rivers as water sources in a speech to the British Association for the Advancement of Science in 1855, and other engineers quickly picked up his refrain.[132] The engineer Thomas Brazill, promoting an immense project to Dublin officials, wrote that "it is now very generally admitted, that all attempts to a supply a large city with water by a series of wells or boring must fail; even when there are basins suited to such undertakings, as at London, Paris, New York, etc."[133]

The shared vision of the enlightened modern engineered water system included as one of its components the requirement that water arrive from a distance. On the one hand, Bateman and his colleagues suggested that cities secure water sources in places distant from populations and from the headwaters of rivers. That water necessarily had to be transported over a distance to the consumers. On the other hand, promoters of a new state of the art held that all towns' water should be delivered under pressure without the aid of steam pumps. Most water companies in the middle of the nineteenth century did not obtain water from upland sources and very few systems had water towers, so companies relied on steam pumping to keep water in their mains under pressure.[134] Most companies could not (or would not, for reasons of economy) keep all branches of their networks pressurized at all periods of the day. Nottingham's water company, created by the renowned Thomas Hawksley, did provide water under constant pressure via steam pumping, and he argued that other companies could also do so.[135] It could be done. But midcentury water company customers could usually rely on flowing taps only during daylight hours or even just small portions of the daytime. By 1890, only around half of the houses supplied by London's water companies had a constant supply.[136]

Bateman and his colleagues, like Chadwick and other reformers, saw a social problem with this old arrangement. The intermittent pumping system, wrote James Simpson, vice president of the Institution of Civil Engineers, was "very inconvenient. . . . Some supplies are given very early, and some very late; an arrangement by which the poorer classes of the community, in particular districts, are often prevented from obtaining water at the time it is most wanted by them."[137] Bateman took obvious pride in eliminating this social defect in his systems. Midway through his career, having constructed twenty-five waterworks, he declared, "I have been instrumental in changing the mode of supply from the intermittent to the constant system."[138]

Bateman and his colleagues advocated large, long-distance schemes because they shared a vision of making rational use of all of Britain's important water sources. They believed that an improved society should distribute water from the regions that had plenty to areas suffering the most want. The engineer Dale Thomas, for example, complained that, in "the Lake districts" of western England, "an abundance of the purest water is flowing year after year into the sea, whilst Towns with dense populations, within a distance of this source of one hundred miles, are suffering the greatest inconvenience of want of pure and wholesome water."[139] An age of progress should reject such limits of deficiency and distance, he argued, and he proposed a pipeline more than four times as long as Glasgow's new thirty-seven-mile aqueduct to redistribute the resource all across northern England.[140] Bateman himself articulated this vision of water organization most comprehensively:

> I never could see the wisdom of the view which would confine the supply of water to the towns or places which lay within any particular watershed. Where the water was most abundant it was generally the least wanted; and towns had grown up where it was often difficult to find or obtain this essential contribution to life and prosperity. . . . As well it might be urged that the coal which is produced in the neighbourhoods of Newcastle should all be consumed in the valley of the Tyne. . . . Water was as much the natural produce of these [Lake District] hills as the artificial products of cotton and woollen in the towns on the plain.[141]

In the hands of the joint-stock companies, water had become a commodity much like any other. In Bateman's view, new water systems should transform it into something as available as air—equally pure, equally plentiful, and equally available to all citizens in all regions.

Growing Towns and the Growing Threat of Fire

Both fundamental and abstract social principles guided the proponents of new systems for providing a regular supply of water for drinking and washing. Proponents also argued for new water systems in order to better fight fires, but those arguments tended to be less lofty. Reformers pointed to deadly and costly disasters inherent with the status quo, the apparent advantages of improved systems for providing water under greater pressure, and the basic need to protect urban society from itself.

Most private water companies in the first half of the nineteenth century provided water only intermittently. An observer considered the textile center of Oldham quite fortunate to enjoy flowing water for five hours per day in 1845.[142] More common was the experience of Rotherham, where water flowed to houses only two hours per day during this period.[143] Also common was the situation in certain parts of London, where water flowed only on alternating days.[144] In the absence of elevated reservoirs or water towers, the water was simply not going to flow into houses of its own accord. It needed to be pumped by steam engines, but while the pumps were filling the town's mains, the flow of water to the service pipes that extended down the streets and to houses was halted.

If a fire broke out at those times, brigades were unable to draw water from fireplugs or hydrants to then pump onto the flames by hand or steam engine. They had to send a runner to turn a series of valves to direct water to the part of town where it was needed. In 1824, bystanders watched hopelessly as a house fire in Edinburgh spread to three neighboring houses while firefighters waited a half hour for the water to reach the scene.[145] In 1839, London firefighters stood by and watched a factory burn and collapse in the hour it took for the water to start flowing.[146] A few years later, London firefighters stood idle for twenty minutes because the valve had been set incorrectly, sending water to the wrong street.[147] An official of the Society for the Protection of Life from Fire estimated that of 838 fires in London in 1849, approximately 550 could have been quickly halted had water been available under pressure near the site of the fire.[148] "Under the present system of supplying water, half an hour transpires on the average . . . before an engine can . . . set to work upon the burning premises," complained metropolitan reformers, "whereas by adopting . . . a constant supply thirty gallons of water per minute can be brought to bear upon the fire in two minutes."[149]

In 1861, the Tooley Street fire in London leveled eleven acres and took the life of a fire brigade chief. Again, no water was at hand because the local water company supplied water to the street for only ninety minutes per day.[150] This disaster led to a House of Commons select committee review and the unification of multiple fire brigades for London into the Metropolitan Fire Brigade under the Metropolitan Board of Works. At the committee hearings, both complaints about the status quo in London and solutions were aired. The existing pattern of intermittent supply to the streets was condemned as "very insufficient" and "very destructive."[151] Witnesses, including an industrialist whose factory burned to the ground before water could be procured, insisted on a constant water supply.[152] And an expert engineering witness called for a centrally administered water supply located at an elevation and maintained under high pressure.[153] Thus, the answer to both the moral threat of insufficient or impure water supplies and the threat of disastrous fires was the same: new, municipalized water systems.

The dual issues of water for fire suppression and water for drinking and washing usually were not linked together in the rhetoric of reformers. That is, most did not describe house fires as a social evil that tended to harm the poor in particular, which was the way that many described poor drinking water supplies—as a social and moral evil. (Henry Mayhew was an exception. He argued that the poor were more seriously affected by fires because they did not have insurance, and even a small blaze was likely to claim all of their property and make them homeless.[154]) But the issues were linked implicitly when activists suggested that poor water supplies for suppressing fires in the crowded parts of the city represented a looming threat to middle-class quarters. It was the same argument invoked when the issue was that poor water supplies in the slums promoted the spread of cholera, which represented a threat to middle-class sections of the city: "There the seeds of infectious diseases are generated, which will spread into other districts where cleanliness is observed."[155]

Investigators described a state of grave "insecurity" in Derby, for example, because water for firefighting was supplied in more established parts of town, but only a few standpipes and a few fireplugs served vast stretches of that mill town, and its tenements did not include protections such as fire-resistant party walls. Any fire, onlookers feared, would necessarily be "extensive."[156] Observers eager to warn the towns of Britain that a fire in one quarter might expand to others were quick to point to a devastating fire that leveled as much as a quarter of Hamburg's central district in 1842.[157]

The Scope of Central State Support for Water Modernization

New, large-scale water systems of course came with equally large costs. Dundee's new waterworks cost £326,000, Bateman's Glasgow scheme totaled £1 million, and Manchester's came to more than £2 million. The average price cities paid to buy out commercial water suppliers between 1840 and 1880 was around £566,000.[158] These were extraordinary expenditures, considering that an average-sized town like Exeter might budget a mere £12,500 a year for public health administration during that period.[159] Cities' funding opportunities were limited. The sole option was usually to borrow money and then service the debt through an increase in the rate burden on citizens. Glasgow issued municipal stock and charged an extra rate against rentals to service the interest on the debt.[160] Manchester increased the existing Poor Law rate by three pennies on the pound, and Cardiff levied a charge equal to between 5 and 6 percent of a property's potential rental value.[161]

The strong predilection toward economy common among elected officials of the mid-Victorian period was the major obstacle faced by water municipalizers. Before long, parsimonious townspeople could be seen cinching their purse strings upon seeing Bateman and his ilk approach. In Glasgow, a town councilor named Gemmel warned that building Bateman's scheme would lead to the "bankruptcy" of the city.[162] When the northern town of Barnsley, with a population of no more than thirty thousand, hired Bateman to submit a water plan, he quoted a project cost of £41,000. "In employing such a person as the Loch Katrine engineer," complained one townsperson, "they could scarcely do otherwise than expect an extravagant scheme."[163] In Liverpool, the proposal of the massive, £2 million Vyrnwy gravitation scheme gave rise to a series of scathing attacks in the local newspaper denouncing excessive municipal spending and subsequent tax increases.[164] Some municipalizers countered that cities could go on charging the same amount for water that the companies had levied, while, of course, providing better service.[165] Others claimed that the income from special water rates could fund further municipal improvements. Some were overly optimistic; other towns, like Leicester, were proven correct.[166]

So, for most of the nineteenth century, opposition to the municipalization and waterworks construction movement tended to be based on the principle of the tight pocketbook rather than on political or ideological grounds.[167] Tory politicians could be found spearheading such initiatives, and Whig newspaper editors could be found lamenting their extravagance.[168] If the development had anything like a partisan basis, it could be argued that water modernizers tended to be the middle-class political newcomers welcomed to urban positions of power by the Municipal Corporations Act of 1835.[169] Having replaced unrepresentative patricians held over from the pre–Reform Act period, these men tended to be anti-oligarchy, antimonopoly, and open to the idea that municipal institutions could be used to elevate the condition of the working class as well as the credit (in all senses of the word) of their towns.[170] Joseph Chamberlain, the great reforming mayor of Birmingham, provides the most famous model of this figure. The mayors of Birmingham and Leicester were hardly confiscatory socialists or even Chadwickean centralizers. Municipalizers were willing to pay dearly for the water companies that they purchased. The middle-class councilors had abundant respect for private property rights. And, while Chadwick had suffered intense opposition from those who viewed his short-lived central Board of Health as an insidious panopticon, the municipalization movement was certainly not one driven from the center.

Municipalities could not look to the national government for much support at all, in fact. Parliament had no intention of giving grants or otherwise subsidizing even a portion of new waterworks. Such a move would have been far too interventionist for the political and economic climate. The government did, at least, give its consent to municipalization in the Public Health Act of 1848, when it formalized towns' rights to pursue the purchase of waterworks. The Public Works Loans Act of 1875 was a slightly more positive step on the part of Parliament to facilitate municipal construction of waterworks and other infrastructure. It provided for low-interest loans from the treasury to municipalities.[171] Practically speaking, Parliament was involved in the majority of waterworks transfers since it required that cities secure authorizing legislation whenever they sought to force the sale of a local water company to the municipality or potentially infringe on existing rights and prerogatives. Town councils frequently presented highly involved bills seeking authority to purchase vast watersheds, cross jurisdictions with aqueducts, issue stock, and so on. But Parliament restricted its involvement to ensuring that cities' projects were not overreaching and that companies and stockholders—and many legislators were water company stockholders—were fairly compensated for their property.[172]

While resistance to municipalization sprang mainly from parsimony at different levels of government, there were other opponents and other grounds for opposition. A correspondent writing to the *Leeds Mercury* suggested that reformers were interlopers and pests who "allowed the city no peace" until it municipalized and improved its water supply.[173] The *Liverpool Mercury* complained that as "the modern water supply system has done away with public wells," it was doing away with community focal points, timeless customs, and sources of ancient magic.[174] Others complained that new water systems advanced town interests over country and showed subservience to labor interests.[175] Though small in scale, such resistance was real. The municipalization movement was not unchallenged.

In 1888, Liverpool completed its own colossal gravitation project, the Vyrnwy scheme, creating the largest artificial reservoir in Britain or Europe at the time.[176] The project flooded a long and deep valley and included construction of a mammoth masonry dam spanning the dale at a height of seventy-nine feet. The project drastically transformed a Welsh region: where once there was the modest river Vyrnwy, there was now a hybrid reservoir-lake spanning one square mile; where there was once a village called Llanwddyn, there was now a terrain whose sole purpose was to serve a city thirty-eight miles distant. This environment of utility most evidently represented the implementation of Liverpool's practical objectives. The thirteen million gallons of water that flowed from the reservoir to Liverpool each day satisfied the town council's goals of freedom from water shortages that had plagued the city in years past and seemed to ensure Liverpool's ongoing prosperity and capacity for future industrial and population growth.[177]

Less evident but no less influential were the underlying values that the Vyrnwy project applied to the land. These were the ideals held by more and more local officials in towns throughout Britain since the maturation of the Industrial Revolution and the population growth and concentration that followed. Correct the problems that citizens living under adverse conditions cannot possibly correct for themselves, was the credo of would-be "moral governors." The credo further called for municipalities to take responsibility for the quality of the urban environment, a responsibility joint-stock companies had shirked, because the hands-off approach of former times would lead to the immediate suffering of the poor and ultimately the failure of society as a whole. Vyrnwy was the application of not only a moral vision but also a vision of the future, a prospect of British cities freed from the bounds of environmental limits. British civilization was the most limitless, advanced, and powerful in the world; the leaders of Liverpool would not watch New York or Paris—or Manchester or Glasgow, for that matter—surpass them in displaying their commitment to the future.

Vyrnwy was also the expression of the values of the British engineering community of the era. Hawksley, who had served on the Health of Towns Commission that helped instigate the water municipalization movement, participated in designing the system and oversaw its construction. Liverpool hired Bateman to evaluate the design before construction began.[178] The scheme was everything Britain's engineering elite judged superior in a water project. It was massive enough to ensure inexhaustible sufficiency, it was situated at a great distance from significant populations, it drew from the sorts of sources of which the engineers approved, and it arrived at Liverpool under pressure. It also represented the organization of Britain's water sources in a rational way, delivering water from where it could be found in surplus to a location where there was great need.

The urban water reformation, to which the Vyrnwy scheme was the greatest monument of its day, was a disjointed movement, not a modernizing campaign from the center. The town council of Liverpool, after all, solicited the plan, hired the engineer, and financed its construction. Here, then, is a more complex picture than historian James C. Scott offers with his examples of central state schemes for modernizing society.[179] In the case of Victorian Britain, local governments undertook projects for the improvement of society, with the environment of the hinterlands serving as the medium for that transformation. The central government, though, gave its explicit acquiescence, if not its tacit support for such activity, in the case of Liverpool, through an act of 1880 authorizing the Vyrnwy scheme. Parliament never denied a local government the authority to undertake such a project on the grounds of principle, only on the basis of practical economic or material obstacles—at least, until the end of the century.

CHAPTER 2

Great Expectations

The First Efforts to Reform London
with Water

"THE PRESENT ARRANGEMENTS of water supply are defective in 'plenty, purity, pressure, and price,'" wrote one commentator regarding London in 1849.[1] Another added, "We ought, by this time, to have learned that the very foundation of moral training in a London tenement is a pipe of wholesome water from the top to the bottom of the house."[2] For London's would-be reformers, just as it was for the campaigners of Glasgow, Manchester, and Liverpool, social reform was synonymous with urban environmental reform. In the same years in which Glaswegians stood in long lines to draw water from a scattering of pumps and citizens of York suffered muddy tap water, Londoners complained of "water-companies who give us bad water in a bad way."[3] And, at the same time that the activists of Bradford argued that the progress of "public morality" could be measured by the extent to which local government took responsibility for putting water in the hands of the poor, so did many in London equate "sanitary ameliora-tion" with "moral progress."[4]

The reformist ideals expressed in hinterland towns by so many were also voiced in London, and the pressures so many towns experi-enced were even more strongly felt in London. Between 1800 and 1880, London added around 44,000 inhabitants per year, or 3.5 million new residents. Its growth rate was greater than that of most other towns and on a far greater scale.[5] None of the rapidly growing industrial cities such as Manchester or Birmingham ever grew to more than approximately 1.0 million persons.[6] The river that London straddled and on which it depended for its water supply was at least as polluted by industry and sewage as Glasgow's Clyde or Manchester's Irwell. And when cholera invaded Britain in 1831, 1848, 1853, and 1865, London suffered more deaths than any other town in the kingdom, approximately 20, 25, 55, and 40 percent of all cholera deaths in those respective years.[7]

The first British towns to take control of their water supplies and build new, improved waterworks began their efforts in 1840, but fifty years later London still had done nothing. The experiences of London offer a sharp contrast to those of other towns. Hinterland towns rejected the idea of improving supplies by regulating privately owned water companies and instead took over the companies' operations, while in London's case, Parliament made attempts at regulation. Whereas in the hinterlands, local water sources were routinely abandoned after being deemed unbefitting an improved community, a royal commission pronounced the Thames a suitable water source for London. And while provincial centers employed John Frederic La Trobe-Bateman and other prominent engineers to create often monumental waterworks, a royal commission rejected a colossal scheme that Bateman offered to London.

Outside of London, local governments transformed water systems in the name of modernizing their towns—with the simple acquiescence of Parliament through easily passed private acts. Because London lacked a municipal government structure parallel to that of Manchester, Liverpool, or Birmingham, for example, the capital could look only to Parliament for help with its water crisis. The ancient City of London, which made up only a fraction of the area and population of the metropolis, enjoyed a relatively effective administration, but the area beyond the limits of its medieval walls—more than a hundred square miles by the mid-nineteenth century—did not. The most basic functions of urban maintenance were carried out with patchy efficacy by parish patricians. Despite suffering the same environmental problems that other industrializing towns experienced, and despite many in London sharing the same ideals of improving lives through environmental reform, the numerous limited local authorities could never hope to accomplish a Loch Katrine or Vyrnwy scheme.

Twice in the second half of the nineteenth century—with the creation of the Metropolitan Board of Works (MBW) and the London County Council (LCC)—Parliament granted London the right to govern itself as a whole and with a degree of real power. When it did, Londoners proved as anxious to reform their environment as the residents of any British city. Indeed, when first granted a relatively central body in the MBW in 1855, London built itself monumental waterworks. But while provincial towns tended to undertake large projects of environmental change for the purpose of water supply, the MBW's major project, the Thames Embankment, was a work for water expulsion. In the eyes of the board, the danger of wastewater retained in the urban environment was more immediate than the danger of poor water supplies. Soon after the Embankment was complete, the MBW developed a plan for a new water source and waterworks to address the secondary danger of insufficient and dirty water supplies. The board, however, was too weak to bring the plan to fruition. It was not apparent that the authority to operate a water supply was within the body's constitution as created by Parliament. Then, in 1888, the MBW was eliminated as part of a local governmental realignment that saw the creation of county councils across England and Wales. To replace the MBW, Parliament created the London County Council, and once again London attempted to achieve the same water reform, through this new body, that many towns had accomplished decades before. And Parliament, though it still tended to refrain from taking a direct hand in water municipalization, showed itself prepared to grant London the same right to take in hand its own water modernization as the government had granted so many times to so many towns. In its first couple of years, the new London County Council needed to secure additional rights and permissions from a slightly grudging Parliament. It also faced wealthy and powerful water companies that, if they were to be forced to sell out, as so many companies had in the preceding decades, naturally wanted to receive as much in return as possible. Despite these obstacles, all signs pointed to London finally joining the water modernization movement in the early 1890s.

The Problem of Water in London in the Early Nineteenth Century

By the mid-nineteenth century, London had a much longer history of private water companies than most towns did. Companies' services were in greater demand from an earlier period because clean water was simply harder to come by in London than in other locations. To begin with, London's focal river, unlike those of most cities, was an estuary, meaning that it was subject to the twice-daily rise and fall of the tides. This tidal action left much of the Thames's shoreline a muddy quagmire for parts of the day, especially in dry seasons, and its water brackish during others, when seawater halted the river's downstream flow. In any case, Londoners had held the quality of the river's water for drinking in doubt since at least the later Middle Ages, and with the passing of the centuries and subsequent increases in shipping, manufacturing, and the population crowding the river's banks, Thames water grew only more unappealing. Several tributaries of the river flowed north and south into the Thames in the region, but these were little more attractive as sources of water. Townspeople and industries had transformed them into sewers before the beginning of the nineteenth century.[8]

Londoners instead drew their water from public or private wells or from public tanks fed by conduits bringing water from beyond the advancing line of suburbs. Those who could afford it might purchase drinking water from water carts. Of the limited supply options, only the water cart's product, usually drawn from a relatively distant and trustworthy source, was appealing as a drink. Beer, rendered sterile during the brewing process and masking the sight and smell of poor water, remained the favorite daily drink of working-class and some middle-class Londoners until the mid-1830s.[9]

From the late 1500s, Londoners who could afford it might enjoy a new service—a piped-in water supply. The London Bridge Water Company, with its huge waterwheel attached to one of the bridge's piers, allowed consumers in proximity to its works to open a tap at their premises and receive a flow of water, thus freeing these townspeople from the need to travel to a possibly distant well and offering great convenience to local industries. Even more fortunate Londoners could enjoy water piped from a dedicated drinking water canal called the "New River," which, beginning in the early seventeenth century, provided clearer water from sources in rural Hertfordshire.[10] By 1655, there were four London water companies providing piped-in water for those who could pay; by 1700, there were six, and by 1800 there were ten.[11] The rate of the companies' multiplication did not match that of Londoners: in 1600, there were an estimated 250,000 Londoners; in 1800, there were a million, with the rate of increase accelerating.[12]

More importantly, the sources of supply available to water companies did not expand with London's increasing population. The Thames remained nearly the only source for the companies. In 1820, five out of ten metropolitan companies drew water from the Thames, three drew from tributaries shortly before they emptied into the Thames, and only two took their water from sources beyond the city.[13] To extract water from the Thames, the companies simply erected either open pipes covered with screens or perforated pipes near the riverbanks.[14]

These rudimentary arrangements invited problems for water company customers. Since most companies drew their supply from the Thames, the lowest point in the city, water would not naturally flow to customers. Water instead had to be lifted into a number of small reservoirs by means of steam pumping so that it would then fill mains through gravity. Because the steam pumps ran on coal, pumping was an expensive operation for the companies. In order to protect profits, companies operated only enough engines to fill a fraction of their network of pipes at a time. On alternating days, one neighborhood would enjoy flowing water, while another would have dry taps. For the same reason of economy, companies never pressurized their pipes at night. Even when taps were flowing, most customers could not expect a heavy flow of water. In the first decades of the nineteenth century, most pipes were still made out of hollow tree trunks, with joints at an average of every nine feet.[15] Companies could not force water through the pipes under much pressure or it would simply leak out of the hundreds of joints interspersed between the pump and customers' taps. Consumers could only hope for a trickle—and hope that their kitchen cistern filled before the company's turncock shut off their neighborhood's water supply for the night.

The companies' method of delivering water directly from the river also invited problems of insufficiency. Without a reservoir positioned between customers and the source, a problem with the pumping apparatus, a clogged intake, or some other malfunction meant that customers immediately experienced stoppages, having no benefit of the buffer a reservoir would have provided.

Supplying water directly from the Thames also raised problems of cleanliness. London did not have a general system of sewers emptying into the Thames in the early nineteenth century, but the river nevertheless had served as the receptacle of stormwater and industrial and household waste for centuries. The volume of this waste only multiplied as London's population reached one million in 1800. Adding to this pollution was the natural silt of a river that drained a watershed that stretched 160 miles into the hinterland upriver from London; winter and spring rains charged the river with muddy runoff. In the first years of the century, Londoners still tended to be more concerned over the outward unpleasantness of their tap water rather than any invisible dangers lurking within. The hard, turbid water made washing textiles difficult, for one thing.[16] The organic material suspended within made the water, especially after sitting for a while, odorous.[17]

The water companies, too, made certain arrangements among themselves that invited problems for water consumers. In the 1810s, the water companies were in competition with one another; between two and five companies struggled to win customers in any given area. Company agents appealed to householders and landlords to switch to their service. Multiple company mains ran down each street, and multiple company service pipes often ran into each building. This was as Parliament wanted. Only Parliament could grant a water company a charter to draw water and sell it for profit, and Parliament granted charters strategically, especially in the early years of the century, to encourage competition. Competition under free trade was supposed to keep prices low and encourage companies to seek out new customers, that is, expand service into new neighborhoods.

As it happened, the practice did result in low water prices, but also very bad service.[18] Companies, nearly bled white by undercutting competitors' prices, could not afford to improve supplies, let alone win investors so that they could spend capital on expanding service into new areas. Facing slow death from competition, the seven water companies that existed in the mid-1810s began to divide London into exclusive service areas; each company, in other words, would enjoy its own local monopoly.[19] The monopolist companies became more attractive to investors, recovered their financial health, and were better able to expand in growing neighborhoods; customers, however, forever lost the ability to switch companies when their service lagged.

Londoners began to look at their water supply with more suspicion in the late 1820s due to several related factors. London's growing population—reaching 1.38 million in 1820—was increasingly depositing its waste into its primary water source; in 1815, for example, Parliament lifted legal constraints barring private houses from connecting to storm drains.[20] There was also an increase in the adoption of the water closet in the 1820s. Water closets were connected to the same drains that emptied into the Thames or one of its tributaries (or they emptied into backyard cesspits, where the effluent frequently seeped into the surrounding soil and groundwater).[21]

Misgivings about London's water supply grew, too, as new literature appeared warning Londoners of the hidden evils in their water. One day in 1827, John Wright, an employee of the radical publisher and politician William Cobbett and a pamphleteer himself, was appalled at the water that came out of his tap. He traced the water from his service pipe back to the main, then to the reservoir, and all the way back to his water company's intake—within a stone's throw of a sewer outfall. He quickly published a pamphlet informing his fellow residents of Westminster that the Grand Junction Water Company was poisoning them.[22] Titled *The Dolphin* after the term for the water intake apparatus, the pamphlet reported that the company sold "a fluid saturated with the impurities of fifty thousand homes [suffused with] animal and vegetable substances in a state of putrification . . . destructive to the health."[23]

In effect, Wright warned that a new era in the history of the Thames had arrived, that the river had become something different than it had been before. The water companies, he argued, were banking on the fact that, to the minds of most Londoners, Thames water remained the merely tolerable stuff of the past. Wright vividly described how it was not.[24] In the following weeks, the *Times* printed testimonials from householders complaining of the fetid smell of their tap water and of "shrimp-like skipping insects" in it.[25] An animated public meeting led by a large group of Lords and MPs passed resolutions condemning the water company and calling for Parliament to investigate the entire metropolitan water supply.[26]

The Royal Commission on Metropolitan Water Supply, prompted by the outcry, agreed that the Thames, "charged with the contents of . . . the refuse of hospitals, slaughterhouses, colour [dye], lead, and soap-works, drug-mills, and manufactories," was being transformed into a social and industrial waste sink, a body of water that could not be relied upon.[27] Many decades before Bateman and other prestigious engineers preached the principle that towns should never draw water from rivers in their proximity, this royal commission concluded in 1828 that "the supply of Water to the Metropolis . . . should be derived from other sources than are now resorted to."[28] And decades even before

FIGURE I. George Cruikshank, *Salus Populi Suprema Lex,* 1832. This etching appeared on a broadside, the publisher of which is unknown. The individual enthroned atop the Southwark Company's intake and crowned with a chamber pot is the company's president, John Edwards. Three individuals are walking on a shoal along the north side of the river, saying, "Devilish thick! Yes, here I stick! It makes me sick!" Along the south side of the river, the individuals are saying, "What torrents of filth come from the Watbrook sewer!! Sewer! why there are 130 such!" Others clamor, "Give us clean water! Give us pure water!"

Edwin Chadwick recommended the same course for London, this royal commission concluded that "the constant and abundant supply of pure Water is an object of vital importance to the inhabitants of this vast Metropolis, that the dispensing of such a necessary of life ought not to be altogether left to the unlimited discretion of companies possessing an exclusive monopoly of that commodity; and that the interests of the Public require, that . . . their proceedings should be subjected to some effective superintendence and control."[29] Just as the Commission for Inquiring into the State of Large Towns and Populous Districts (1844–45) would conclude, the 1828 commission doubted that water companies, by their nature, could put the public welfare before profit, and the companies therefore needed a degree of government supervision.

Parliament did not act on the royal commission's recommendations (though the public incrimination did motivate the water companies to slowly shift to a policy of filtering their supplies through large sand beds).[30] The government, it seems, held too strongly to a belief that only arm's length regulation, if any at all, was appropriate in such matters. In a separate incident earlier in the decade, a parliamentary committee considering limits to water rate increases concluded that it could not "justify an interference of the Legislature affecting private property."[31] The companies' own self-interest, the government believed in both cases, would ensure that they provided customers a good water supply. This was the first of the central government's several attempts at regulating London's water supply during the century—a prominent contrast to the pattern in provincial towns, where the strategy of regulation was consistently passed over in favor of purchase.

Thus, in the 1830s, in the absence of regulation or competition between companies, Londoners still had no recourse when their water company failed them—and fail them the companies did. In 1832, George Cruikshank linked poor Thames-drawn water supplies to the cholera epidemic of autumn 1831; he did not directly indict water in the spread of the disease as Dr. John Snow would in the outbreak ten years later, but he drew the practical conclusion that, without clean water, "we shall all have the cholera."[32] Writing on the health of Londoners in 1837, Dr. John Hogg was scandalized that, though seven years had passed since the royal commission had called for reform, London's water supply remained "deficient for the purposes not only of health, but of comfort and cleanliness."[33]

Calls in London for a Modern-Moral Government

In London, as elsewhere, members of the community's elite castigated a society that would leave its urban environment so dangerously unreformed. For them, the visitations of cholera and the quotidian suffering of the poor in the confines of the urban warrens—both being conditions that, reformers believed, would be eliminated by providing clean, plentiful water—showed how far London was from having a just society. In 1850, the *Edinburgh Review* compared indifference on the part of the governing class with nothing less than the murder of the lower classes, arguing that more than twenty thousand working-class Londoners a year were cut down "by causes which, if we chose, we might expel by a current of water. Though we do not take these persons out of their houses and murder them, we do the same thing in effect—we neglect them in their poisonous homes, and leave them to a lingering death."[34] In 1849, Charles Lushington, the MP from Westminster, and other elites arranged a public meeting to agitate for water reform in the wake of a cholera epidemic, "after so enormous a number of our population having been laid waste by that fearful pestilence." He argued that "the poor . . . had not the means of alleviating their suffering within their houses by the plentiful use of water which would have been a great means of placing a shield between them and certain death."[35] And in 1850, the chemist Arthur Hassall expressed the new ideal of moral government repeated throughout provincial cities in the period, writing that "one of the most important and distinguishing features of the present age is the attention now bestowed on all matters connected with the sanitary condition of the people. . . . To contribute in any degree to the improvement of the social, moral, or sanitary state of the people must be a subject of rejoicing and congratulation alike to the Executive and the individual."[36] For these commentators, reforming London's water supply meant turning away from an era when authorities merely threw up their hands in the face of poor public services and instead turning toward a more active supervision of the urban environment based on their ideas of morality.

Reforming the water supply would also result in reforming the morals of the poorest Londoners. Most critically, countless reformers closely linked impure water supplies to the abuse of alcohol throughout the century. "The universal testimony of our Missionaries is, that a large, a very large amount of drunkenness is occasioned by the great difficulty of obtaining pure water to drink in many of the poor parts of London," wrote Reverend John Garwood of the London City Mission in 1859.[37] "Pure water affected the moral condition of the people," Henry Fawcett stated in the House of Commons, "because, without pure water, it was hopeless to expect temperance."[38] There are too many instances of this kind of argument in the historical record to count.[39]

Improvers believed that without water readily at hand to clean bodies and homes, the character of the London working-class family would be degraded. "There is a most fatal and certain connexion between physical uncleanliness and moral pollution: the condition of the population becomes invariably assimilated to that of their habitation," stated the *Edinburgh Review* in 1850.[40] In 1854, a contributor to Charles Dickens's weekly magazine *Household Words* looked forward to a moral renaissance once water reform prevailed. "Our most pressing concern," wrote John Morley, "will be with water supply and drainage. There must be a constant supply of good water at high pressure within reach of every housewife's thumb." Once fresh water was widely available, put to use in cleaning working-class homes and bodies, and the waste water ejected, the higher qualities of humankind would be liberated. "By the time that is all done, we shall have advanced also in the moral and mental discipline of urban life to a better state," he wrote. "Put London in perfect order as a town most fit to be occupied by living bodies, [and] it will have become also the best place for the health of growing minds and souls."[41]

Pride in their towns and, by extension, British civilization was a critical motivating factor for water reformers in provincial centers, but in London the association was far more immediate. There, where the pressure of astounding population growth combined with extraordinary epidemics made the water crisis graver than anywhere else, the metropolis was looked to as a model at home and abroad. The chemist Arthur Hassall offered a particularly vivid warning that London's dignity was in jeopardy. A visitor from "some remote kingdom," he wrote, would never believe that "in a city . . . inhabited by the wealthiest and most distinguished for their skill in the mechanical arts, a system of water supply prevailed, by which dwellers in that vast city were made to consume their own excrement and offal."[42] In 1849, a pamphleteer complained that Paris could boast of its many public water sources and that New York, with only four hundred thousand inhabitants, had a water supply four times as great as London, with its two million inhabitants. "Why should we not," he asked, "be in every sense the greatest Metropolis of the World?"[43] Dr. John Hogg, too, asked why, when the cities of Persia and Turkey were dotted with public watering spots, "we should be so far behind these less civilized states in the establishment of fountains, baths, and reservoirs?"[44] And the *Edinburgh Review*, perhaps most accusing of all, told its readers that "the shameful condition of our great capital in this respect [of water availability] is now fully before the public; and the credit of England is, we think, involved in the course which we shall at last deliberately take."[45]

Great Expectations for London

For many, the example that London should follow was not only the one offered by New York or Paris but the one already being set by a number of British towns in the 1840s and 1850s. The editor of the *Times* wondered why the first city of the kingdom could not enjoy a constant supply in all quarters, as some provincial cities did. "The thing has actually been done," he wrote, and "it may be seen at Greenock, Glasgow, Paisley, Preston, Sheffield, and many other places besides, and it is found by unmistakable practice that wherever these water supplies thus existed, . . . the cholera was disarmed."[46] Many in London argued that their city should do as Manchester had done in 1846 and Halifax in 1848: eliminate the local water companies. "Any government who would break up the present system of London water supply," wrote Arthur Hassall hopefully, "would be entitled to lasting gratitude, and I am certain would receive it."[47] When concerned West End inhabitants gathered in 1849 to discuss the water problems, they called for "placing the entire control [of the water supply] in the hands of the Inhabitants themselves."[48] The 1844–45 Commission for Inquiring into the State of Large Towns and Populous Districts concluded, too, that the "duty of providing" water should reside in the local administrative body.[49] And the General Board of Health, born in the aftermath of Chadwick's unpleasant revelations and the Public Health Act of 1848, also called for the government takeover of the water companies. They had failed London, the board concluded, especially during the recent cholera attacks. During the epidemic, their focus "was constantly called to the inferior quality and deficient quantity of the water supplied to the Metropolis, as well as to defective distribution . . . and on numerous occasions the effect of polluted waters in causing a dreadful excess of mortality was forcibly brought under their notice."[50] The findings of the board were not legally binding, but they suggested the elimination of the companies and placing control of the water supply in the hands of a single government agency.[51] As elsewhere, the idea that it was the *duty* of government to provide its citizens with the most basic necessity was becoming more common in London.[52]

The belief that cities must draw their supplies from a distance also gained prominence in London at the same time it did elsewhere in Britain. The ascendance of this belief was not inevitable. One alternative was to try to clean rivers that flowed through cities, to prohibit pollution by industries, or to deter sewage pollution. But in a variety of ways, and in the hands of a variety of actors, the idea that cities should reach over and beyond water sources in their neighborhood for unblemished sources in their hinterlands won out.[53] As Manchester began constructing its Longendale waterworks to transport water from twelve miles distant, some called for a gravitation scheme for London to transport water thirty miles, others seventy miles.[54] Some called for reservoirs in the elevated region of the headwaters of the Thames; others called for a series of wells in the sandy soils south and west of the metropolis.[55] Whatever the external water source, reformers expressed the urgency of securing what they considered a pristine supply as quickly as possible.[56]

A Government for the Sake of Water Supply

John Stuart Mill added his voice to those calling for the municipalization of London's water supply in 1851. Water should be in the hands of local government, he wrote, but, "in the case of London, unfortunately, this question is not a practical one. There is no local government of London."[57] The vast majority of the sprawling city was administered on the level of the parish by vestries. The parish was the smallest geographic unit of the Anglican Church, and the vestry was a council of parish fathers—some popularly elected, some restricted to the elite—who saw to such matters as road maintenance and, in particular, poor relief. In addition to the vestries were the various bodies of commissioners, which were of different sizes, authority, and composition and which Parliament had created haphazardly since the mid-eighteenth century; each usually served a single purpose, such as seeing to lighting, road maintenance, or public safety in various limited jurisdictions. The ancient City of London, the roughly square-mile center of a metropolis that covered more than a hundred square miles, enjoyed a relatively effective, if arcane, government, but that body took no interest in administering areas beyond its medieval walls. And the seventy-eight vestries and nearly one hundred commissions were far from the active, effective bodies reformers could count on to administer a modern city of which they could be proud.[58] For J. S. Mill, the inward-looking authorities of the City of London and the decentralized administration of vestries and commissioners—what there was of it—did not qualify as a local government at all, let alone the kind of organization that could operate London's water supply.

The horrors of the cholera epidemics, the shocking mortality rates, the dreadful reports of Chadwick confirmed by the Health of Towns Commission—all of these constituted an argument difficult to ignore. The new nature of the Thames and the new realities of the London environment demanded action on the part of a national government that was averse to regulating trade. In 1852, Parliament required that

London's water companies draw water from the Thames at a point west of the metropolis; thereafter, the companies had to submit to the Board of Trade plans to draw from new sources.[59] In the short term, the 1852 Metropolis Water Act alleviated some concern about dangerous water quality—cholera's threat seemed less close at hand—but worry soon shifted from London's sources of pollution to contaminants from upland areas.[60]

The 1852 act did nothing practical to address reformers' concerns over the lack of water availability. It required companies to offer consumers a constant supply, but a wide loophole allowed them to avoid this obligation. The companies had to supply water continuously only if four-fifths of occupiers of the companies' respective service areas applied in writing; given the economic and educational status of a large segment of London's population, the stipulation incapacitated the act. The 1852 act, in sum, is not remembered as a turning point in water reform.[61]

In the same years, though, a change in the administration of London's environment was slowly developing. The national government began to take London's drainage problem more seriously. In 1847, Chadwick and his allies succeeded in securing a new body to look at London's drainage on a large scale: a special commission with authority to investigate and partially direct the sewerage of the metropolitan area. It was underfunded and lacked much authority, but it began the work of addressing half of the two-part predicament Chadwick described: not enough clean water entered the urban environment and not enough contaminated water exited it. The commission, renewed roughly annually between 1848 and 1855, succeeded in eliminating thousands of cesspools. These sumps had kept dangerous waste in proximity to living and working quarters, and many were suspected of leaking into the surrounding soil. The commissions also required that thousands of buildings connect to local sewers.

Some commentators were uneasy about replacing thousands of cesspools with a single cesspool in the form of the Thames; even Chadwick himself hoped it was only a temporary expedient until the government could find a permanent repository for the waste.[62] However, according to miasma theory, then the most commonly held theory of disease transmission, it was imperative to submerge waste underwater as quickly as possible to avoid the spread of infection. Many thought that miasmas, invisible but stinking clouds of pollution or poison, either spread disease directly or made people predisposed to it; the theory seemed commonsensical given that disease and odor were often found in the same quarters. Put waste immediately underwater, believed individuals like Chadwick, and no miasma formed.[63] The commissions came to the conclusion that a unified sewer network, with sewer mains of great capacity, was the only way to ensure the comprehensive drainage of the metropolis. Society could be better protected from the attacks of epidemic and the disgrace of filth if the subterranean world of London conformed to a systematic drainage regime.

The cholera epidemic of 1854 helped to elicit action. One powerful MP, Benjamin Hall, a political opponent of both Chadwick and his vision for the central administration of water reform, was overwhelmed by cholera's devastation, and he turned his substantial political skill to the problem. There must be some wide-ranging sanitary administration, an overhaul to the "state of this vast city." He said that "unless great and speedy radical changes in the constitution of its local affairs [are] effected, it [is] utterly hopeless to expect those affairs to be well conducted."[64] He suggested creating a congress of representatives sent from the vestries.[65] The new Metropolitan Board of Works he envisioned represented a change in the status quo, a concentration of power to some degree, but it was also an essentially practical response to London's inability to take action given its fragmented geography of jurisdictions. Hall was no radical, but he was motivated; he wanted to be sure that his new board had the authority to take in hand its life-or-death directive: eliminate polluted water from London. It was not democratic, consisting of delegates from vestries that were either selected by a limited electorate in some cases and literally handpicked from a restricted clique in others. Nor was the MBW to be a central health agency, like one envisioned by Chadwick, or a central government for the metropolis.

The defenders of the ideal of government noninterference protested even this limited concentration of power, but, though vocal, they were few in number, because cholera tended to shout down resisters. In 1855, Parliament secured the Metropolis Management Act, which finally created Hall's unified board, one responsible for the basic duties formerly performed by scores of small authorities spread across the broad face of the metropolis.[66]

A Monumental Work for Water Expulsion Rather Than Supply

The forty-four members of the Metropolitan Board of Works, appointed by their home vestries, immediately set to their most important duty of quickly removing waste from proximity with London houses. They hired a former engineer to the defunct sewerage commissions, Joseph Bazalgette, and a platoon of surveyors, draftsmen, and clerks. Bazalgette had already been involved in the construction of many London sewer lines, was thoroughly familiar with numerous proposals for a universal London sewer system, and came fully endorsed by the engineering establishment.[67] Within a year, he offered a plan for eighty-three miles of sewers—really brick tunnels—to collect the discharge of thousands of drainpipes. These would intercept the contents of existing drains and sewer lines running toward the river and shunt all of their contents away to riverside outfalls east of the city. The most critical intercepting tunnels would border the Thames itself, like a last line of defense. It took some years for Bazalgette, the MBW, and the government to settle on a plan that was inexpensive enough to satisfy a parsimonious government while at the same time extending the sewer mains far enough eastward to alleviate fears that the pollution would be carried back to London on the rising tides.[68] But the MBW, aided by an intolerably hot July that lowered the level of the Thames until it consisted of 20 percent sewage and gave rise to the legendary "Great Stink of London," finally received approval for its plan from Benjamin Disraeli's Tory government and Parliament in 1858.[69]

In creating the MBW and authorizing this important project, Parliament was hardly committing itself to taking a strong hand in a wider urban reformation. Benjamin Hall and other supporters of a cohesive works board for London were intent on giving such a body authority over various jurisdictions so that it could take unified action, but at the same time they remained committed to a liberal ideal of low taxation and low government expenditures. The MBW, in other words, was given a clear mandate but was saddled with a bit of a handicap: the act that created the board empowered it to take only a small portion of existing vestry rates. In order to pay for its extraordinary project, the board had to plead for low-interest loans against future rates from the Bank of England in order to cover contractors' large initial expenses. In the end, the amount lent to the board was barely enough to fund most of Bazalgette's new sewer system.[70] The most visible and arguably most important part of the system, the riverside intercepting sewers, would be encased in new river embankments. The cost estimate was no less than £3 million, and the national government, other than providing loans at 3.75 percent, was not going to contribute anything to these works. It was only by a stroke of luck that the funds of a tariff—of medieval origin—collected on coal and wine imported into the city became available at this key juncture.[71] Since the duties were an established tax on Londoners and would be used to fund a London project, Parliament allowed their £1 million reserve and £160,000 annual interest to fund the construction of the Thames Embankment.[72] In 1869, engineers completed the Albert Embankment (named after the Prince Consort) across from the Houses of Parliament, they completed the Victoria Embankment (in which a novel subterranean railway ran) on the opposite side of the Thames in 1870, and, in 1874, the Chelsea Embankment, built on either side of the river in west London, was completed.

London's Thames Embankment, and the system of drainage it represented and crowned, belonged in the category of great waterworks that conspicuously demonstrated so many towns' self-proclaimed modernization in the period. It stood as evidence that, given a local authority to take such a project in hand, Londoners, too, would eagerly transform their environment, changing London's riverside from a malleable thing of soil to a regularized thing of concrete, from a space ruled by tides, to one ordered by engineers. The embankments' invisible heart is part of what made them a modernizing technology in the eyes of urban environmental reformers; at their core were the great tunnels that collected the contents of thousands of lesser sewers, six hundred thousand tons of sewage and wastewater per year, and bore it east of the city to outlets on the north and south banks of the river.[73] There, great pumping stations—again, ornate, grand constructions—helped maintain the eastward flow and lifted the water into holding reservoirs until it could be released (untreated) into the river on the outgoing tide.[74]

The Thames Embankment project was a work that was supposed to modernize London by flushing clean the darkest slums of the city while at the same time putting a majestic outward face on the river that had formerly served as an open sewer. The embankments stretched for a total of five and a half miles, stood thirty feet high, and were faced with hundreds of thousands of square feet of imported granite.[75] By encasing the river shore, they eliminated the banks of mud and reek-

ck that emerged at low tide.[76] They were topped with trees,
vide roads, and pedestrian walkways. A member of the Royal
Society for Arts ascribed to the Thames Embankment a civilizing effect, anticipating that it would not only serve for "the purification of
the river" but would also lead to a "respectable state of its shores."[77]
The Embankment appealed to British pride by making the Thames
respectable, too, in the eyes of foreign visitors; upon the completion of
the various sections, the MBW proudly claimed that they "were works
mainly for the purpose of large improvements in the metropolis, and
to make the metropolis of London a city that would compete with cities on the Continent."[78] And a popular London guidebook stated that,
"while the Seine at Paris, a far inferior stream to the Thames, contrib-
utes one of the most beautiful features to the French metropolis, the
Londoners have hitherto persisted in shutting out from sight their far
more magnificent river and converting its stream into a sewer."[79] But
the guidebook author was happy to relay that, with the Metropolitan
Board of Works having taken things in hand, the bank of the river "is
now converted into a magnificent promenade."[80]

A Project Equivalent to the Embankment for Water Supply?

Surely, believed many Londoners, action to address the second half
of the water reform challenge would follow closely behind the success of the Embankment. Bazalgette and the MBW had eliminated an
immediate danger, to the minds of Londoners, by quickly submerging waste underwater and shunting it far east of the city, but that left
the second danger: polluted water supplies. "The water has been . . .
positively filthy to drink," complained one consumer to the *Times* in
1874. "Perhaps when some fever has set in in the neighbourhood . . .
the managers will turn their attention to the matter," agreed another.[81]
Reformers also were not satisfied with either the volume of water entering London or its availability. A vestryman from St. Martin in the
Field complained in 1879 that London still lacked a sufficient "supply
of pure and wholesome water" distributed throughout the metropolis
to ensure "proper conditions of safety, equal to those enjoyed by large
provincial cities and towns."[82]

The body that served as a local government for London had taken
responsibility for discharging foul water, and now many looked to it
to take control of providing clean water, as other local authorities had
done. Observe, argued one Southwark medical officer, "the number

of most important and successful instances in which the water supply has been lodged with the representatives of the people."[83] He concluded that "as soon as possible the water supply must be taken out of the hands of the trading companies altogether. Their functions . . . in this respect are an anomaly. . . . It is a barbarism, and behind the age. London in this matter ought to without delay be served at least as well as the larger provincial town and cities."[84] Those who supported municipalization were bolstered by yet another royal commission on London's water supply, meeting between 1866 and 1869. The government convened the commission, explained one commentator, in response to "a popular impression not only that the water actually furnished was impure . . . but that it was objectionable in principle to supply, for human consumption, water drawn from large rivers draining the country through which they flowed."[85] In 1869, the metropolis's population stood at three million, twice as high as it had been forty years earlier; meanwhile, water demand per capita had almost doubled.[86] Doubts about the inexhaustibility of the Thames grew into "the prevalent opinion with the general public," reported one journal.[87] There was pressure to find an alternative source of drinking water, and this issue raised questions about who was responsible for securing the supply. The water companies showed no interest in exploring for other sources. New works required great outlays of capital, and there was no competition to motivate improvements in service.

The last thing the water companies were likely to do was construct a massive gravitation scheme on the model set by cities like Glasgow or Manchester. But a water scheme of the modern fashion is exactly what John Frederic La Trobe-Bateman—now sixty years old, a thirty-year veteran of waterworks engineering, and a fellow of the Royal Society—offered to the 1869 commission and to London.[88] His plan was to transport water from the valleys of Wales to London. The reengineered landscape he described to the royal commission embodied all of the principles he and his fellow leading engineers advocated to urban modernizers. First, it rejected the traditional boundaries for water extraction. The supply of London ought to be sought where the water was purest, softest, and most abundant, and "most secure from injury by any operations of manufacture or agriculture," he told the commissioners.[89] If pure water happened to be locked up in the northern Wales, so be it; Bateman eyed a watershed in a region south of Mount Snowdon and to the east of the peaks of Plynlimmon and Cader Idris, the headwaters of the river Severn.[90] Next, the reformed landscape must provide for the superabundance that freed a community from any concern about future growth. Bateman's plan called for the construction of reservoirs that would gather the waters of existing lakes and streams and hold several months' reserve supply. The system, he told the royal commission, would be capable of delivering an amount equal to that which London drew from the Thames each day.[91] And, as important in the eyes of Bateman, the water must arrive at the city from its gathering network under high pressure. An aqueduct (chiefly a canal, but tunneling through hills and crossing valleys in pipes when necessary) would bear the water across 170 miles before it emptied into one large reservoir situated on a hill ten miles northwest of London; the hilltop reservoir would stand 250 feet above the metropolis.[92] The project, in sum, embodied the paradigm he and other eminent engineers had been making manifest in the secondary cities of the kingdom, but it exceeded them all in ambition.[93] Like a master describing his intended magnum opus, Bateman gave nineteen hours of testimony on the plan before the royal commission and was said to have done so without referring to his notes or data.[94]

But after meeting for two and a half years, the commission reported that the Thames was a perfectly suitable source for the future supply of the metropolis. Cholera had beset London as the commission began meeting in 1866, and the outbreak focused on the area served by the East London Water Company in particular, but the commissioners nevertheless certified the purity of the drinking water offered by the companies.[95] The main expert witness on water safety analysis, the chemist Edward Frankland, tried and failed to show how cholera spread in water, let alone to show the commissioners that London water was dangerous to public health, which he desired to do.[96] The commissioners went on to express confidence that the Thames would provide sufficient water for any demand the metropolis was likely to make of it in the future.[97] And, of course, the commission balked at the enormous expense of the project, even though they were convinced of the Welsh water's high quality.[98]

Naturally, the water companies emphasized the aspects of the royal commission report that cast them in a good light. The report supported their argument that "most people greatly exaggerated" the amount of "excrementitious matter" in the river Thames, they claimed. Besides, they wrote, "the germs, the fungi, the cells, or whatever is supposed to propagate the disease [cholera] are only imaginary existences."[99] And the companies argued that the commission praised the companies' ongoing efforts to expand continuous supply service. Since Bateman's Welsh scheme might have spelled the end of the profitable companies, they seemed relieved that the commissioners raised doubts about the necessity of the monumental project.[100]

Despite their optimistic depiction of London's water situation, the commissioners saw the municipalization of the companies as critical. "The future control of the water supply should be entrusted to a responsible public body," the commission concluded. The central administration of the water supply, it reported, "offers the only feasible means of introducing efficiently the system of constant supply, and for securing a compulsory supply to the poor." It would make water cheaper, more pure, and more available for fighting fires.[101] Commercial companies could not be relied on to ensure that water was close at hand in all areas in the modern city, nor could they provide the ubiquity of supply that modernizers needed if they hoped to compel working-class Londoners to change their practices.

The 1869 report of the Royal Commission on Water Supply rebuffed Bateman's monumental waterworks scheme for London, but its endorsement of municipalization encouraged the MBW to pursue further the question of London's water supply—with an importation scheme designed by Joseph Bazalgette as the centerpiece of its strategy. "The Metropolitan Board of Works is the proper body," said one member in MBW debate, "to whom the water supply can be intrusted. A sufficient proof of our capabilities is afforded by what has already been done in draining the whole of London and carrying out the Thames Embankments."[102] Board members cited the public health risk of intermittent supplies, the enormous benefits to firefighting that a high-pressure supply would bring, and, once again, the "good example afforded by Manchester and other provincial towns."[103] The MBW took steps to bring about the same two-part plan towns like Manchester had enacted numerous times: purchase the local water company, then modernize the city's waterworks.

Another ineffectual attempt at limited regulation on the part of Parliament, though, delayed action. Acting in response to the 1869 royal commission that adamantly insisted on the value of a constant water supply, Parliament in 1871 secured an act meant to require a high-pressure supply from the water companies. But, after resistance from the well-funded, well-connected companies, Parliament allowed them to name the conditions under which they could not be compelled to do so—a right of which they made great use, to the point of practically nullifying the act.[104] In the absence of any real reform, many vestries and district boards, especially in less affluent areas of the metropolis, appealed to the MBW to eliminate the companies, citing in particular the problems of poor water quality and arbitrary fees.[105] By 1877, the MBW was prepared to present two bills in Parliament: one for the purchase of all of London's water companies, and one to authorize Joseph Bazalgette's scheme to provide London with a new water source.

Bazalgette, aided by two respected consultants, devised a plan to import water from a region approximately twenty miles south and west of London.[106] Adhering to the principles advocated by engineers for decades, Bazalgette rejected the water of the neighborhood river; instead, he would draw London's supply from subterranean chalk formations in the hills of Surrey. Unlike Bateman's gravitation scheme, the chalk-drawn water plan would require pumping stations situated in the water-bearing areas to force the water to the surface and up into four covered reservoirs placed on hills north and south of London. Descending hundreds of feet to the city below, the water would arrive under pressure, ensuring a constant supply to every corner and elevation of the metropolis. Microscopic examination and other tests, the MBW's engineers reported, showed the water to be impeccable, while a recent examination of the companies' water had recently revealed "moving organisms."[107]

Bazalgette's chalk water scheme represented an engineering of compromise that mirrored the nature of the MBW as a body of compromise. Parliament had not endowed the board with the authority of a true town or borough council, but it was still expected to address the problems of the urban environment as a true borough council would. The water scheme was not designed to replace completely the Thames water supply. London would still rely on Thames water for manufacturing, domestic washing, street cleaning, as well as water closet, drain, and sewer flushing, and so on. The new chalk-drawn water was meant only for drinking, food preparation, and, since it was a high-pressure supply, fire hydrants and plugs. The companies' existing mains and service pipes would remain unchanged, with the MBW laying new conduits for its dedicated potable water, with fire hydrants regularly spaced along the new mains.[108] The chalk water scheme would provide only 13 percent of the total amount of water demanded daily by the metropolis.[109] It was to cost around £5.5 million, compared to around £22 million for Bateman's monumental scheme (which included purchasing the water companies' existing works).[110]

Bazalgette's water proposal failed. In the eyes of the MBW, it appeared well received, at least "on the part of the organs of public opinion."[111] And, indeed, the *Times,* which in later years would thoroughly revise its stance on municipalization and waterworks construction, gave the Bazalgette scheme a positive verdict on the grounds of high water quality, high water pressure, and affordability.[112] But opposition outside the walls of Parliament increased, with some members of the MBW blaming the hostility of powerful water company interests.[113] Within Parliament, the companies' supporters came to their defense. "I do not think . . . that the Water Companies can be fairly charged with not having fulfilled the duties imposed upon them," said Joseph Samuda, an MP for Tavistock in the West Country; their efforts, he contended, were satisfactory in light of the challenges they faced in serving such a large population, and the quality of water they supplied was, at least, sufficient.[114] Besides, the complaint continued, the MBW's plan for replacing the drinking water supply was absurd: "a more crude, undigested, and impracticable scheme has never been put before the public."[115] The approbations of Bazalgette, who had been knighted for having directed the creation of the Thames Embankment, seemed quickly forgotten.

The water scheme's detractors chiefly opposed it by linking it with the bill under consideration in Parliament to allow the MBW to purchase the chalk water sources. The board was unsuited to running a consolidated water supply serving hundreds of square miles, opponents claimed; board members should "confine themselves to the purposes for which they were appointed."[116] The provincial towns that had municipalized their water supplies had had the benefit of being, as town or borough councils, undisputed central local authorities. The MBW could be viewed as little more than a particularly large commission for sewers, streets, and lighting. The water companies' sympathizers promoted this view, and even MPs sympathetic to water reform for London could offer no very strong ground for claiming the MBW was on par with a town council. Henry Fawcett, Liberal MP for Hackney, argued that the MBW, "as far as water Bills were concerned, [was] responsible for the government of the Metropolis, [if not] then what other municipal body was?"[117] Another MP argued for placing the water supply under the control of one public body, but in the same breath he admitted that "whether that body should be the Metropolitan Board of Works was another matter."[118] The purchase and waterworks construction bills did not proceed far.

And so London's case remained as a peculiar exception within the larger development of water modernization in Britain. Not only had London remained outside of the water modernization movement for decades but once it received something like a local authority, it settled on an atypical strategy. The urban water reform movement was marked by bold rejections of the status quo and visionary plans to build massive waterworks to provide for countless generations. The limitations of local resource availability, the challenges of distance and geography— none of these were to impede the modern city. In Bazalgette's water supply scheme, London settled on the "engineering of the possible." One anonymous letter writer to the *Times* chided London for accepting Bazalgette's modest scheme in 1877, stating that "it will indeed be a matter of surprise to the country if the Metropolitan Board seriously entertain . . . a scheme that will give paltry dribbles of hard water, measured out through pin holes, while Manchester and Liverpool are seriously considering schemes to augment their already excellent supplies from distances of 100 miles. I trust that Londoners will show themselves to be animated by the same spirit as has been displayed in other large cities of England."[119] The limited measure proposed by the compromise body would have been a far cry from those of Manchester or Liverpool, created by Bateman and like-minded engineers. Bazalgette proved himself the master of water drainage engineering, but Bateman's prescription for town water supply remained dominant.

A New Government for London, a New Government of Water?

In order for London to build a water supply system of the Bateman model, the city would require, it seemed, a full town council. The home secretary, H. A. Bruce, argued early on that "in order to give effect to the recommendations of the [1869] Commission it seems necessary to create a central authority for the metropolis. . . . This measure can not be properly carried into effect unless in conjunction with . . . creating a general government for the metropolis."[120] And already in 1862, a very active and reform-minded Westminster vestryman, James Beal, was contending that "Sir B. Hall's [Metropolis Management] Act was but a bridge between bad government and municipal action. It has

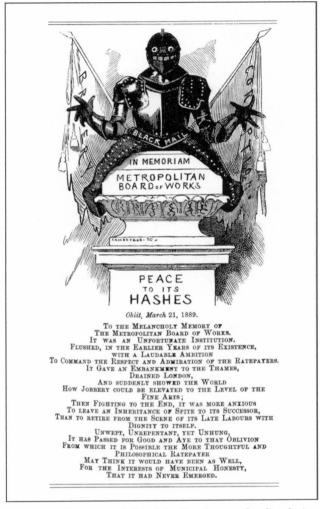

FIGURE 2. Cartoon published in *Punch* upon the dissolution
of the Metropolitan Board of Works, March 1889.

served its purpose."[121] For Beal, acquiring London's water supply would
require a "Metropolitan Municipal Corporation" to "make the Metropolis as powerful as Liverpool, Glasgow, or Dublin."[122] Given such a central government, "Parliament will grant to London what it has already
granted to Manchester, Leeds, Liverpool, Edinburgh, and other great

towns."[123] And Joseph F. B. Firth, an equally active London reformer, wrote in a treatise describing a future government for London that its water supply must be transferred to a public body but "certainly not to the Metropolitan Board of Works; a body created for a different, and almost temporary purpose."[124] For Firth, too, a single, popularly elected "Supreme Council" was the only body capable of "acting fully" to provide London with a constant supply of clean water.[125]

Within a decade of the failure of its water bills in Parliament, the MBW obliged critics like Beal and Firth by dissolving under a cloud of corruption. In an age that had seen the slow but steady expansion of the electorate, the quite indirectly representative MBW was especially vulnerable to charges of irresponsibility when prominent members fell under suspicion of financial misconduct and favoritism in property transactions in the late 1880s. When the body raised taxes over the course of time, critics charged that any taxing authority should be directly answerable to the taxpayers.[126] The board, after all comprised delegates from the vestries, most of which were elected by male electors who satisfied certain property requirements. And one-third of all London vestries consisted of members simply handpicked by preceding vestries. It also hardly helped the MBW's reputation that it held its most important meetings in secret.[127]

Unrelated to the disgrace and collapse of the Metropolitan Board of Works, Parliament was, in the MBW's final years, reviewing the issue of local government throughout England and Wales.[128] The central government sought to increase the authority of local government at the county level so that local taxes could fund local projects and so that more issues generally would be handled on the regional level instead of being referred to Parliament for adjudication. Naturally, London, which sprawled across parts of Middlesex, Surrey, and Kent, presented a dilemma. Should its administration be divided among those counties, should the management of London be split between a new metropolitan body and those counties, or should Parliament invent a county where none had ever existed—a County of London?[129] In the end, the Conservative government decided to create a London county council at the same time that it created county councils throughout England and Wales through the Local Government Act of 1888.

The new London County Council consisted of 118 members (plus aldermen) sent from the same constituencies that had members of Parliament, two from each to serve three years. The electorate that sent county councilors, though, was wider than that electing MPs; it included all adult male householders and even female heads of household, who comprised no less than one in six of all LCC electors.[130] The new body took over all of the MBW's responsibilities: sewerage, fire prevention, parks maintenance, streets and bridges, and building codes. However, the act that created the London County Council did not formulate new powers for the body in order to make it more effective than the MBW had been; indeed, the LCC was saddled with the vestries—which jealously protected their prerogatives—as a second-tier government. Unlike the MBW, though, the new county council's mandate was clearly to serve the huge area of the metropolis as a central authority, rather than to serve as a meeting of vestries temporarily united to solve problems deemed otherwise intractable.

Here, finally, was the body to take in hand London's water modernization. So believed James Beal, who at age sixty had retired as a vestryman but returned to politics when the LCC was created, convinced that it was the body that would enact water reform. The new council would immediately take up the matter, he declared, and he predicted a rather straightforward success.[131] He ran for a seat on a platform of the elimination of the companies and won.[132] The London County Council took its first steps toward municipalization and waterworks modernization within weeks of its inaugural meeting. The early meetings had largely been devoted to constitutional and organizational issues; one of the first policy initiatives the LCC took was a resolution to form a special subcommittee devoted to water supply purchase and new water sources.[133]

In moving to take the first steps toward water reform, the councilors cited the "scandalous" state of water supply in London, but within the council and within the wider public discourse, the need for change and the expectation of change was widely felt, so long had the water companies' sources and manner of delivery been implicated. In addition to the attention brought by the 1828 and 1869 commissions, the 1874 Royal Commission on the Pollution of Rivers concluded that the Thames should be abandoned as a source of drinking water. In 1876, a parliamentary select committee on fire prevention had concluded that the water supply should be operated by a single public body in the name of safety, and, citing public impatience for action, the House of Commons appointed another select committee in 1880 to research buying out the companies, though no action was taken.[134] Already, a

Times editorial had stated in 1879 that "we are all agreed, as we have been for thirty years[,] that the water supply of London is bad. We have talked about the matter and inquired into it over and over again, and yet nothing has been done."[135] In 1890, the *Times* offered the now-familiar litany of reasons to support municipalization, from doubts that the companies would be able to provide for an expanding population, to the belief that commercial companies could not build a modern gravitation scheme, to worries that the companies were not safeguarding supplies against pollution, to the simple principle that the prime necessity of life should not be in private hands.[136] While "investigating committees" had long concluded in favor of central control of the water supply, a writer on municipal issues in 1890 stated that "until now the suitable authority has not existed." The new county council, "encouraged by the vestries and all London," was now demanding to take its rightful place in charge of the water supply, and "the demand cannot be long resisted."[137] Now that London had "a representative council," the *Financial News* expected municipalization to follow as if by logical progression. "There ought not to be any difference of opinion," it stated, since "the principle which underlies [municipalization] has received acceptance in several of the great provincial centres."[138]

James Beal, who since the 1860s had watched towns like Leeds and Edinburgh municipalize their water supplies and build modern waterworks while his calls for water reform in London went unrealized, became chair of the London County Council's water committee. His committee, unsurprisingly, did not even debate the appropriateness of municipalization. It was not only Beal who believed it necessary to modernize London's water supply; most of the committee's members simply echoed the sentiments and arguments that foregoing urban reformers in Birmingham, Glasgow, and Leicester had rallied to for decades.[139] The economist and statesman Sir Thomas Farrar, for example, believed that London could become a "model city" or realize an "improved civilisation" only if the LCC operated London's water supply in the public interest.[140] Frederic Harrison wrote that "the natural business of the local bodies" was to provide "pure, unlimited gratuitous water, which stands on the same footing as air."[141] The committee immediately began to research methods for purchasing London's companies by inquiring into the experiences of provincial towns that had municipalized and by calculating the stock exchange value of London's

eight companies.[142] Later, Beal sent letters to the eight companies solic-
iting their official positions on their willingness to sell.[143]

Just as London's would-be "water reformers" echoed the arguments
of past reformers, so too did the Londoners take up their established
course of action. The LCC's water committee researched securing a
new water source for London at the same time it investigated municipalization.[144] Most towns developed new water sources and works soon
after purchasing their local water companies, and the bills they sent
to Parliament seeking means to municipalize often also requested authorization to buy land and secure rights to new water sources and
fund new works. Part of the LCC water committee's mandate included
investigating new water sources, and the committee, in turn, directed
the council's engineer to research both the capacity of the Thames to
continue supplying London's growing population and alternative water sources. The engineer investigated other towns' water-importation
schemes, hired rainfall experts to study different watersheds, and com-
municated with individuals who studied London's regional water table
to explore the possibility of a well-drawn supply along the line of Bazalgette's, among other possibilities.[145]

The goals of the LCC's water committee were encouraged by the
majority of the council, with little regard for party affiliation—just as
had largely been the case in provincial towns. Municipalization's supporters included conservative Liberal Unionists like Sir John Lubbock
and orthodox, cost-conscious Liberals such as Sir Thomas Farrer, while
most Tory-allied councilors were not wholly committed one way or
another.[146] Many of the metropolitan vestries, too, threw their support
behind both the LCC's takeover of the companies as well as the council's early research into new water supplies and waterworks.[147] And the
water reform efforts of the new county council still enjoyed the favor
of "the organs of public opinion."[148] But when James Beal died in the
summer of 1891, pursuing the municipalization of London's water supply until his final days, the LCC water committee's progress toward
water reform was not equal to the effort it expended or the confidence
it inspired.

The new London County Council turned out not to be the picture-
perfect body to enact water reform that it had appeared to be at first
sight. The Local Government Act of 1888 withheld from the London
council an important handful of powers it granted to all other coun-

ties. The council had no power to police, it had to submit a budget or money bill to Parliament annually, and it had no innate authority to purchase and operate public utilities. It was not that the Conservative government that created the LCC suspected that it would be a stronghold of its political rivals; the Conservatives had won forty-five out of fifty-nine London parliamentary seats in the previous election.[149] Instead, the checks on the LCC's powers had to do with a long-standing general jealousy with which successive national governments tended to regard London.

This animosity seemed to stretch back centuries, and at least somewhere near its root were the age-old wealth and arrogance of the City and perhaps the unruliness of the London masses.[150] The Tory prime minister at the time of the LCC's creation, the Marquis of Salisbury, embodied this hostility in his own way, regarding London as the entirely hopeless mass surrounding Parliament and the queen—the only worthwhile things in the expanse. He preferred not to be there.[151] As far back as the 1860s, James Beal blamed in large part a general suspicion about the metropolis for Parliament's hesitation to grant London a more powerful government.[152] Parliament, too, was simply cautious in its creation of a new authority presiding over the largest population and enjoying the most valuable taxable property in the kingdom, save the central government.[153] According to Asa Briggs, "The leader of [the new council's] dominant party, Conservative critics of their own government feared, might become more powerful than 'any prime ministers and some monarchs.'"[154]

So, while Beal and the LCC's water committee were determined to purchase the water companies from the first moment, they had to wait to receive Parliament's official sanction to enter into negotiations with the companies in 1890.[155] But the London County Council found all of the companies unwilling to negotiate. According to them, the LCC's concurrent investigations into alternative water sources, potential rival sources to the Thames, would lessen the value of their undertakings.[156] The water companies could wait for good terms; they were so successful that they could pay dividends averaging around 9 percent in the 1890s.[157] The following year, yet another parliamentary committee threw its support behind municipalizing the water supply, recommending that the London County Council promptly purchase the companies by arriving at a fair price with them or, following a precedent often repeated by towns that had previously municipalized, through entering binding arbitration under a government-appointed arbitrator.[158]

While the LCC's water committee was pleased that the principle of water supply on a commercial basis was once again being criticized, it was not satisfied with the recommendation to enter swiftly into arbitration. Quite simply, by putting its fate in the hands of an arbitrator, the council could be committing itself to an excessive price, a precedent the water committee had seen in previous cases of municipalization.[159] First, the water committee wanted to know the exact capacity of the companies to provide sufficient water for London's expanding population in the future and also the companies' ability to protect their water from poison. The LCC would hardly be modernizing if it simply provided the same poor service in the future that the companies had provided in the past. And if the water companies were not up to the task, as the water committee expected they were not, the committee wanted the purchase price to reflect this liability, so that the money saved could be invested in improving the water system.[160] Second, the LCC needed to know more about the companies' capacities in order to calculate exactly how much additional water the metropolis would require from a new hinterland water source. The body could not compel the water companies to reveal their capacities, so it needed the government to demand the evidence. So, bolstered by the show of confidence from an 1891 select committee, the LCC asked for a royal commission to examine the suitability of the Thames as a water source and the ability of the water companies to serve London in the future. The government agreed, and the inaugural session of the London County Council came to a close while the new Royal Commission on Metropolitan Water Supply was meeting.[161]

In the provincial cities, the forces of industrialization led to urban expansion, and at the same time, certain concrete responses—and ideals of moral governance—of reformers expanded the power of local government. Reforming the environment to preserve the new city, in other words, demanded new centralized powers. And the growing towns won that power time and again, with Parliament, having ensured that water company stockholders were quite well compensated, proceeding to sanction purchases, establish new authorities, and finance often-mammoth projects of environmental organization.

London was subject to the same pressures as industrializing provincial towns. As elsewhere, a quickly growing population concentrated itself in a relatively small area in order to join together in manufactories and participate in the building of the city. In 1800, approximately 1.0 million, in 1820, about 1.4 million, and in 1830, 1.65 million Londoners lived shoulder to shoulder on the banks of the Thames so that by 1830 the river became "saturated with the impurities of fifty thousand homes." But still Londoners relied on that river for their water supply. From the 1820s, Londoners called public meetings, wrote letters to the *Times,* and petitioned Parliament to require that the water companies provide cleaner water. In subsequent decades, concerns about water's superficial qualities turned into grave anxiety about the connection between poor-quality water and disease. After the 1831 epidemic, individuals like George Cruikshank pointed to links between cholera and the water supply. After the 1848 incursion of cholera, Dr. John Snow's confirmation of a definite correlation was slow to become widely known, but after the 1854 epidemic—with London suffering more than half the country's fatalities—individuals like Sir Benjamin Hall needed no more convincing of a connection, even if its precise character was unclear.

The absence of sufficient water input and output through the metropolis provoked moral outrage in London, as had been the case elsewhere. The wealth and power of London alone did not signal an advanced civilization. Only when the wealth and power were harnessed to reform the urban environment for the benefit of Londoners of the lowest station would philanthropists be satisfied. Like the reformers of Bradford who called poor water arrangements a "flagrant social crime," those humanitarian Londoners believed the degradation of the urban environment must equate to the degradation of human morals. There could be no moral governance on the one hand against many hundreds of thousands of unwashed rooms, more unwashed bodies, and a filthy river on the other. Establishing a morally governed city would necessitate a properly governed environment, and the engineering culture volunteered to do just that. The very champion of the movement, J. F. Bateman, laid a plan for a monumental water system—the largest and most conspicuously modern plan yet devised—at London's feet. Waterworks engineers throughout Britain followed Bateman's largely self-formulated ideal principles, which called for high water pressure, abandoning local rivers as sources, and importing water from isolated sources in the hinterland, and he meant to apply these principles to London, too.

The case of London's attempts at water reform in this period reaffirms that "modernization" or "civilization" through environmental reform was the work of multiple, disconnected nuclei rather than that of the single central state. Subject to these pressing influences and ideals, the towns of the hinterland took environmental reform into their own hands. They did not wait for a directive from the national government—none was coming. The perceived crises were uniquely local. Local governments committed themselves to purchasing water companies, building new waterworks, and accepting the financial and administrative burden of governing their water supply. Because of the special conditions of London's growth, with a very high rate of expansion around a core—the ancient City of London—that was uninterested in governing areas outside of its walls, the metropolis lacked a local government like those Manchester or Birmingham had gained in the first half of the century.

So the national government was the only authority that could take in hand London environmental reform; it had not dictated to the provinces in the matter and only reluctantly directed London. Only when it was widely agreed that there was a real crisis did Parliament, under an odorous cloud from the river, deign to invent the Metropolitan Board of Works to accomplish what separate sewerage commissions could not: the Main Drainage and Thames Embankment. The central government had no taste for giving this body expanded responsibility, in the form of water supply control; instead, it tried to regulate the London water companies through parliamentary acts in 1852 and 1872—acts that proved toothless. Provincial towns did not even attempt regulation, instead moving straight to municipalization. And, quite simply, the national government proved its lack of interest in the urban water modernization movement when it left the power to operate public utilities out of the London County Council's constitution in 1888. Parliament, at least, proved more interested in safeguarding against an overly wealthy and influential regime in the LCC than solving London's water question.

Through 1891, then, London's water companies continued to enjoy the status quo in spite of the hopes of so many Londoners who deplored their service, of so many Londoners who saw them as obstacles to advancing civilization, and in spite of the powerful trend that saw water companies' elimination up and down the country. If there was an expansion of state intervention in Britain as the twentieth century approached, this case certainly does not show it.[162] It instead offers an example of a government that only indifferently regarded the expansion of London central government and only impassively came to the verge of granting London's government the power to municipalize the local water companies, which many town councils had won generations before.

When the second London County Council was returned in 1892, the nature of London's water supply question was transformed. In the eyes of the central government, the 1892 election confirmed disturbing trends in London politics with the addition of new councilors with new aspirations for London. That caucus brought about a view of water supply and waterworks even more hopeful than that of the water reformers of the past, who had seen, in the introduction of water into the urban environment, the "progress of true civilization." That new group of county councilors brought to their office a dedication to transforming the urban and hinterland environments on a scale even exceeding the works of Glasgow and Liverpool; its intentions so amplified the water reform movement that the central government reconsidered its long-standing accepting stance toward the modernization movements centered on British cities.

CHAPTER 3

"Communism in Water"

A Strategy for Harnessing Water to Reshape Late Victorian London Society

O N THE EVE of the twentieth century, London's government attempted a radical program of urban reformation. Its plan comprised a re-theorization of the city itself—a new concept of the relationship between people, the environment, urban technologies, and local government. London's reformers believed that, as it was, the metropolis was a concentrated but disjointed and chaotic amalgamation of souls. Until then, the role of government had been to wade into the chaos and marshal what little order was possible and to carve a little space for equality or community out of the free-for-all of individualism. In its new plan put forward in the early 1890s, London's government, the London County Council, sought to build—slowly, from the ground up—a city that generated and reproduced communitarianism and egalitarianism automatically.

Slow and from the ground up—or, rather, from beneath the ground: the first target of change was London's water supply. The issue of water seemed mundane. A previous generation of urban officials had established water supply as the fit and proper domain of municipal government. This humdrum veneer made water the ideal vehicle for initiating a "subterranean" revolution. In the London County Council's plan, the average Londoner was to have no idea that the city was undergoing fundamental transformations. On the face of it, the plan was simply to alter the way that water was collected and delivered. First, London's government intended to buy out and eliminate the metropolis's private water companies. It had every reason to expect that the move would be popular because the city's water suppliers had a reputation for providing poor service and water of dubious quality. And the move would be uncontroversial because many British cities had taken the same step of water municipalization in previous years. After eliminating the companies, the council planned to overhaul the water collection and delivery system, abandoning old sources and works for state-of-the-art ones. This the council could justify on the ostensible grounds of improved purity, ubiquity of service, fire safety, and, again, the precedents in other cities.

Considered in isolation, these steps hardly appeared the stuff of revolution, but London's plans for water did not exist in isolation. The new water system was meant to be integrated into a wider program of changes in urban technologies and institutions and policies. At the same time that water supply was to be reordered and retasked in the service of new principles, London's government planned on transforming how Londoners lived, traveled, learned, and worked. Water supply was to be just the first, though critical, component of a far-reaching program of reordering the urban environment, transforming institutions, and changing the workings of the capital in the name of social justice. London pursued a vision of technical systems for public health that focused on reshaping ingrained social patterns—not just ameliorating immediate dangers to society. The initiative represented a rejection of a limited program of local government—one that provided only certain "necessities," narrowly defined—in favor of a broad interpretation of what was right and practicable.[1]

A close look at the new strategy for water in London provides an opportunity to grasp how components of nature could be rallied, mustered alongside institutions and policies and technologies, in a governmental complex or regime aimed at molding society.[2] Thus, this case shows particularly clearly how actors hoped that water could be employed as an apparatus, closely linked to others, for realizing social transformation—not just more authority or profit for the local authority or safeguards for public health. Students of the Victorian urban revolution have underestimated the importance of water in town councils' strategies for building and sustaining the new liberal city. Worse still, they have overlooked the London County Council's plans to use water to realize more fundamental social change than the builders of the Aswan dam or the irrigation works of the Ganges plain had envisioned.

Another main point of recovering this history is to challenge the idea that it was inevitable that public health technologies like water supply systems were divorced from the matters of justice and equality in the nineteenth century. As history played out, Victorian public health institutions and engineers focused on making a regime of industrial exploitation more "survivable" for the working class—just survivable enough to get working-class males into and through a reasonably healthy adulthood of demanding labor. But the path toward a merely palliative public health infrastructure and administration was not preordained.[3] There were those who fought against Chadwick in order to place poverty and injustice at the root of health problems in industrial cities. Individuals like the influential doctor and reformer William Pulteney Alison argued that government should intervene to prevent poverty in order reduce disease, instead of merely removing poisons from the midst of urban crowds. Address the disease, they argued, not mere symptoms. Though Chadwick's circumscribed vision of public health succeeded, it did not have to.[4]

London's effort to reconceive and redesign water supply as part of the means of tackling disease itself and not merely the problems of urban society is one that, like Alison's argument, undermines a sense of inevitability about how the difficult situation arose in the first place. It resurrects a moment when public health and the city itself could have come to mean something new. And as fantastic as its plans to transform the city sound, the London County Council came close to succeeding in its plans for water supply—laying at least the cornerstone of the wider program of which water supply was integral. In a different British city, in the hands of a different local body with a different political profile and reputation, water could have surreptitiously begun to serve the mission that London envisioned for it. But, as it was, the plan drew strident opposition on ideological grounds once its full significance became clear. In no other city had water municipalization been stopped dead by the national government, but the promoters of the water plan were obvious in their ambitions for society. The conservative opponents of London's reformation managed to redirect the flow of water toward serving their own vision.

Ultimately, the London County Council's reconception of the role of water supply in urban society succeeded in transforming politics and society in the city—not because London's government succeeded, however, but because it failed. The opponents' victory set London's water history on a course that continues to determine the relationship between Londoners, water resources, and their government.

Enter the Fabians

Between the 1840s and 1880s, more than 150 local governments assumed responsibility for providing their towns with water. They either took over existing water companies—usually greatly expanding their waterworks soon thereafter—or built new water systems themselves. First the government of Manchester, then Leeds, then Glasgow, then Birmingham added water to the list of municipal duties until, by the 1880s, more than 80 percent of Britain's town governments operated a water supply. In part, officials took this step because they considered it the appropriate response to immediate health threats. The period saw steady urbanization and industrialization; water for drinking and washing became harder for city dwellers to obtain even as industrial pollutants and human poisons threatened its cleanliness.

But interpreting the sweeping movement to municipalize water supply as an unthinking reaction to new threats to health obscures the wider intentions that urban officials identified with regard to water. For them, water was a tool for realizing an urban society that was industrious, expansive, and moral; in other words, a good water supply could generate a liberal Victorian society. Municipalizers wanted water supplies to protect working-class populations, especially from epidemics that would incapacitate them; if workers were sick, they were not working and keeping the wheels of commerce turning. At the same time, philanthropists taught that disease directly attacked the moral strength of the nation when it decapitated the working-class family, removing the breadwinner and sending the surviving family members into the poorhouse. Many municipalizers also argued that they wanted to use water to attack the demoralizing power of what they euphemistically termed "filth." In their understanding, working-class townspeople became degraded in the face of grime and smells that they could not escape in the absence of plentiful clean water. This degradation led to further poor habits, sin, and drink, which would propel the poor into a downward spiral. Thus, water had the special ability to program morals even as it freed city dwellers to live the life of the modern liberal urban civilization. In sum, while it is true that town councilors often framed their support for municipalization in terms of defending public health, in words and deed they revealed that they saw water as a governmental tool for realizing a wider set of conditions that historians associate with liberal Victorian Britain.

In sum, urban leaders had absorbed water into a wider Liberal governmental regime for decades before the socialistic Fabian Society arrived on the scene. Water was supposed to help preserve and encourage large and relatively healthy family units and, by extension, large, dense urban populations. New supplies were also aimed at serving industry directly and preventing fires. And waterworks would scrub the face of urban Britain clean, so that the country could present to the rest of the world a face of which Britain could be proud and thus impress the world with its successes in achieving profit, strength, freedom. Liberal local governments took water supply into their hands for the purpose of realizing that core vision of Victorian liberalism. But once members of the Fabian Society won influence over London's government, they sought to radically re-task water to serve a very different socialist governmental system never before pursued.

Sanitarians had spoken of the health of the individual working body; London reformers spoke about the health of the social body of workers, the unitary social organism. Sanitarians acted with a sense of urgency and were often reacting to imminent crises such as cholera epidemics; London representatives acted on a broad vision of fundamental change. The sanitarians and philanthropists chiefly sought to transform local living conditions, to halt visible suffering. If they had ambitious, visionary moments, if they pictured a sort of Chadwickean ideal city, they dreamed of Britain's largest cities free from disease, industrial society safe from ruin and even made robust thanks to the health of its fundamental units—the laborers. They envisioned shiny faces and cities to rival those of America or France.[5] Those guiding London's policy did not seek reform for reform's sake. Though they might have mentioned water quality and sufficiency in their literature, they were not primarily motivated by the desire to offer the poor safer drink; they hoped that, by making London a communal machine, by making it a machine of democratic socialism—they would transform society. The London County Council's vision for water—as a component of an integrated socialist governmental regime—extended far beyond the sanitarians' vision.

Articulating that vision for the LCC's majority party was the Fabian Society, a small organization located on the political spectrum somewhere between the Marx-influenced Socialist-Democrats on the left and the far more conventional London Liberals to the right. Rather than a party, the Fabians were a group of earnest thinkers, writers, and politicos who aimed to eliminate an economic system that, to their minds, precluded the possibility of living a moral life. They believed that the poor could not have justice and the middle class—from which the Fabians were largely drawn—could not be absolved of guilt for their complicity in economic injustice until the exploitative and even brutal structures and practices of capitalism were eradicated.

London received a metropolis-wide representative government for the first time only in 1889 as part of a parliamentary act that created new county councils across England and Wales. It was at that moment that the Fabian Society found an ideal focus for their plans for social transformation. London's and other county councils would be manageable units for the group's members to try to influence and ideal bodies through which to legislate and enact their brand of socialistic change. Annie Besant, a women's rights activist and by far the most famous Fabian of the late 1880s, expressed her hopes for the new councils in sanguine terms: "The division of the country into clearly defined areas, each with its elected authority, is essential to any effective scheme of organisation. . . . In perfect unconsciousness of the nature of his act, Mr. Ritchie [Tory president of the Local Government Board] has established the Commune. He has . . . created the machinery without which Socialism was impracticable."[6] George Bernard Shaw and Graham Wallas, both Fabians, were optimistic that the new county councils could acquire the power to own land or means of production, and they thus believed that the councils would be the perfect units of government to turn into units of ownership.[7] On the eve of the creation of the first LCC, Annie Besant wrote that "the *mot d'ordre* for Socialists now is, 'Convert the electors; and capture the County Councils.'"[8]

By 1892, Sidney Webb, who was the chief Fabian theorist, and a number of fellow Fabians sat on the London County Council, and they immediately set about persuading their colleagues to legislate a new urban society. Their numbers on the council were small, but the Fabian Society literally authored much of the platform of the LCC's majority coalition of Liberals, Radicals, and labor representatives, collectively known as the Progressives.[9] The Fabians envisioned themselves as the operators, the LCC as its control board, and the institutions and tech-nologies of the city as the machinery for generating a communitarian society. The plan was to use the LCC to provide Londoners with hous-ing and employment, to order living spaces, industrial spaces, and rec-reational spaces along communitarian standards, to build up municipal services and industries and dismantle individualistic, profit-oriented ones. In return for this guardianship, London's citizens would give back funds in the form of taxes and payments for goods and services provided by municipal enterprises, returning wealth to the common municipal purse. Within several years the London machine would pro-duce a generation of Londoners who could not remember a time before they looked to the community—to the commune—for their livelihoods, health, and happiness. Such a machine was to enjoy a continuous en-ergy loop; it would provide for Londoners, who, in turn, would provide power and resources to the socialistic machine.

The Fabians published *The London Programme,* their platform for the Progressives, the same year they took their seats on the council.[10] Webb announced with enthusiasm that the program "is based on the idea of a complete revolution in the administration of nearly every department of London municipal affairs."[11] It described an LCC with authority unprecedented in the history of British local government. The council would have new regulatory powers and offer novel ser-vices in addition to absorbing many of the responsibilities of the forty local parish vestries and district boards.[12] The LCC of the new Lon-don was to comprise a new market authority, oversee the creation of public hospitals, and provide a unified public welfare system to replace the disconnected Poor Law providers across London's dozens of local authorities.[13] The LCC would also have new powers for "Land Munic-ipalization," that is, authority to commandeer land for the purpose of building improvements, clearing slums, and so on.[14] The ideal LCC was to be one of the kingdom's largest employers and would use this dominance to influence labor practices in favor of workers. Under LCC control, workdays would be reduced and pay increased.[15]

For Webb, these administrative changes in London governance were to serve shoulder to shoulder with physical ones in the transfor-mation of society. Among other projects, the ideal LCC was to build workers' housing in a circular zone orbiting London, construct radial tramway lines connecting them with the center, and provide free tram service to the workers. A property tax levied on the landlords who otherwise would have lodged those workers in poor inner-city housing would fund this system.[16] In this plan, the LCC would shelter its citizens, convey them to and from their workplaces, and extract the means to do so from the workers' would-be landlords. The LCC of the new London was also to buy the metropolis's three private gas companies in order to provide Londoners better service than the profit-seeking companies. As it was, the gas companies neglected poorer neighborhoods because connecting to tenements promised little business while posing extra financial risk. The LCC, lacking a profit motive, could provide gas service for dwellings housing the poor while also illuminating dangerous stairways, slum areas, and alleys.[17]

The difficulty for the Fabians was in initiating their reforms. To do so, they had to overcome not only those in favor of dominant laissez-faire economics but also those with an economic interest in maintaining private city services and utilities. But Webb and the Fabians had a solution, a fuel that would quietly but effectively start their machine in motion: water. Operating London's water supply on a municipalized basis was to be the Fabians' simple method of introducing socialism to London without the ideologically opposed, self-interested, or simply conservative-minded citizenry taking notice. In the 1890s, eight profit-seeking water companies provided Londoners with their water; the LCC simply had to win parliamentary approval for purchasing the companies at a fair value and they would take the first step toward the commune.

A New Theory of Water Supply

Because of the governmental strategy of which it was a part and the rationales on which it was based, the act of municipalization in London would be more than the mere repetition of a step taken by scores of towns and cities already. One of these new rationales was Webb's historicist outlook, which suggested to him that great civilizations— societies that subordinated the individual for the sake of the health of the whole community—satisfied the basic utilitarian needs of the community first and with conspicuous completeness. The plan was also based on the Fabians' vision of the total rationalization of the country's natural resources, especially for the sake of providing cities in need with resources locked up in the less-needy hinterlands. And the plan was based on a strategy of having the municipality employ as many of its citizens as possible. Middlemen would be cut out. Laborers would work for fair wages and reasonable hours. Most important, capturing the water supply was based on the belief that landlords, in this case "water lords," were the root of much exploitation and of the continuance of an unjust cycle that ensured the poor would remain poor and the wealthy would act as an unproductive sponge off the majority of the community.

In *The London Programme, Fabian Essays in Socialism,* the Fabian Tracts, and other writings, the Fabians explained how capturing the water supply would aid in the slow regeneration of society.[18] The project appealed to some first and foremost for the way it served the Fabians' campaign against rent.[19] Rent was at the root of all social evils, according to Bernard Shaw and many of his fellow Fabians. It was rent that allowed a fortunate minority of the people to live off of the toil of the miserable majority of the people. "The income of a private proprietor can be distinguished by the fact that he obtains it unconditionally and gratuitously by private right against the public weal," wrote Shaw.[20] And as the nonlandowners worked the soil, or improved economies, or flooded the cities to drive up property values, the property owners could charge more rent without having done any more work.

Water rates were just another form of rent—a price paid by the public to the privileged minority who, "simply because they are sons of their fathers," somehow managed to come into possession of water sources, along with their abetting shareholders.[21] When water rates rose, they did so because population growth pushed up the rental value of the houses on which the water rate was based. "The New River Company's Water Shares have their present value, not because [the original seventeenth-century company founder] Sir Hugh Myddleton's venture was costly, but because London has become great," wrote Sydney Olivier, a founding member of the Fabian Society.[22] The rent, too, was simply extortionate because of the companies' and shareholders' greed. In his *Figures for Londoners,* Webb claimed that London's "water lords" charged 41 percent more for their water than it cost the companies to supply it.[23] He wrote that it cost less than £700,000 a year to supply London with water but that Londoners paid £1.7 million for it.[24]

And what, asked the Fabians, gave that privileged minority the right to tax others for the use of a part of nature in the first place? Another Fabian, the Reverend Stewart Headlam, condemned landlords who claimed "the seashore and the rivers; so that . . . every salmon which comes up from the sea might just as well have a label on it, 'Lord or Lady So-and-So, with God Almighty's compliments."[25] And Webb protested that "our landlords steal from us even the Thames . . . its industrial advantage [goes] to swell the compulsory tribute of London's annual rental."[26]

In an ideal London in which individual owners no longer possessed the land and, in this case, water, the public would pay a tax to itself for their use. Instead of a water rate paid to one of the eight private suppliers, the water rate would be subsumed under a general tax. "The existing 'water-rate,' equalised and properly graduated, might continue to be levied as part of the County Council rate; but there is no reason why any special charge should be made for water, any more for roads," wrote Webb. "We can, at least, [have] 'Communism in water.'"[27] And when the profit motive was removed from the provision of water, the Fabians suggested, Londoners would enjoy important water services that had formerly been too costly for the companies to provide.[28] The companies were hesitant to provide water to the tops of tall tenements— usually home to the poorest residents—because of the added pumping costs, among other reasons. Webb envisioned the LCC providing water to greater heights: "We see in imagination . . . the County Council's mains furnishing, without special charge, a constant supply up to the top of every house."[29]

Income like that earned by the municipality of London for providing its people with water, wrote Shaw, "must always be held as common or social wealth, and used, as the revenues raised by taxes are now used, for public purposes."[30] Extending to the public management and services would provide income for further extending public management and services. Bernard Shaw envisioned "an annually growing fund" for London's "improvement."[31] Webb wanted water supply proceeds to contribute to social welfare, explaining that, if London itself provided public services, "it might save at least £1,500,000 every year— enough to cover half the expenditure on the relief of London's poor."[32] And Webb wrote that the LCC's ability to secure an inexpensive loan for the purchase of the water companies and the "saving likely to accrue from unification of management [of the water supply] would amply suffice to provide any improved service required, as well as afford a useful surplus towards the cost of London government."[33] Water was just one component of a wider, integrated system of health and equity.

The Fabians' vision of centralizing London's water supply was also a component of a much broader vision of making rational and efficient use of the whole country's wealth and natural resources. This reflected the Fabians' utilitarian inheritance; it also reflected the illiberal side of Fabian socialism. Enlightened administrators at the center would divide the products of the land and nation to provide the greatest happiness to the greatest number. Bernard Shaw saw this enlightened redivision at the heart of the socialist enterprise, and, as only he could, he cast it in lyrical terms. He described a growing desire among common people that the "gifts of Nature might be intercepted by some agency having the power and the goodwill to distribute them justly. . . . This desire is Socialism."[34] One aspect of resource rationalization was the logical distribution of resources across the face of the kingdom; thus, resources like water supply would not only be stripped from private hands but intelligently distributed. Annie Besant described the redivision of resources in less poetic terms: "the Central National Council . . . [would effect] the 'nationalisation' of any special natural resources . . . enjoyed by exceptionally well situated Communes [to her mind the progeny of the county councils]."[35] Why should a geographical region with abundant supplies of a universal necessity enjoy a monopoly? asked Graham Wallas. "Those forms of natural wealth which are the necessities of the whole nation and the monopolies of certain districts, mines for instance, or harbors, or sources of water supply, must be 'nationalised,'" he wrote in *Fabian Essays in Socialism*.[36] Applying this principle, Sidney Webb proposed that London import water from the rain-soaked hills of Wales to give the metropolis a superabundant, unpolluted water supply.[37]

The pursuit of the water supply was particularly suited to Webb's historicist way of thinking. His conception of the future evolution of English society toward socialism was based on what he perceived to be a historical progression from hyperindividualism toward collectivism. His thoughts about the future progress of London, particularly in the area of water supply, were colored by what he regarded as the collectivist and utilitarian example of imperial Rome. In "Rome: A Sermon in Sociology," Webb wrote that the British imperial center should follow the example of the Roman imperial center—or rather Webb's interpretation of the Roman administration, for Webb's Romans had a decided scent of socialism, Benthamite utilitarianism, and Spencerianism. For example, the Romans, wrote Webb, were supremely community minded and rejected the individualism that might otherwise have eroded their strength. He wrote that "the essential point which the

Romans never overlooked, is to maintain at all costs the paramountcy of the social over the individual interests."[38] In a Spencerian frame of mind, Webb compared individualist Athens to utilitarian Rome, writing, "Rome seized the higher truth which Plato dimly saw [in *The Republic*], that natural selection now operates on communities more than individual men, and Greece fell because [they were a] comparatively inferior community."[39] As part of that rejection of individual interests, Webb's Romans were devoted to the powerful central government and its institutions. "There appears not the faintest desire for any limitations of the power of the Government; no Liberty and Property Defence League," he wrote. Instead, there was a "grand tide of devotion to an Ideal City."[40] In pining for his imagined golden age that would have no calls for limits on government, Webb reveals the illiberal strain of Fabianism.

His Romans focused on building the basic infrastructure and the spectacular projects that lent strength to the social organism. The building of temples, wrote historian Webb, was a much lower priority than roads and aqueducts.[41] In supporting the project to import water from Wales, the Fabians wanted Londoners emulate Roman engineers; "aqueducts larger than Rome ever contemplated," Webb argued in *The London Programme*, "must be undertaken for the city whose size and whose wealth Rome itself never approached."[42] And in *Facts for Londoners*, Webb asked, "Would it not be well for London to emulate ancient Rome, and allow its millions unlimited opportunity to wash? Communism in baths, as in roads and bridges, . . . could not have other than beneficial consequences on the public health."[43]

The recent history of successful municipalization of water supply across Britain also furthered the Fabians' focus on public access to water. For the Fabians, that history of municipalization demonstrated that a sort of "unconscious socialism" was slowly gaining ground.[44] As evidence of this, Webb pointed out that almost half of the gas consumers in Britain received that fuel from public authorities and that sixty-five local authorities borrowed money for waterworks or municipalization in 1887 and 1888.[45] The Fabians also saw a historical tendency toward democracy in England, significant because socialism, they believed, was the natural product of democracy. "Always and everywhere democracy holds Socialism in its womb," wrote Hubert Bland in *Fabian Essays in Socialism*.[46] Shaw cited as evidence the growth of local, central democracies like the London County Council—the "local machin-

FIGURE 3. "The Triumph of New London," from *New London: Her Parliament and Its Work* (London: Daily Chronicle, 1895), a work reprinted from a series that originally appeared in the *Daily Chronicle*.

ery" for replacing private enterprise with state enterprise.[47] And just as local democracies would replace profit-seeking services and industry, anticipated Webb, public services would be "democratized" as another step toward the social-democratic "Ideal City."[48] Webb rejected the idea of an unelected board, an oligarchical group like the Thames Conservancy, overseeing the water supply. Not a "nominated Water Trust," but "a Statutory Committee of the London County Council" should operate it, he argued.[49] London's electors—soon, Bernard Shaw believed, to be every adult regardless of income or even sex—were to control their own land, resources, services, and industry through such LCC committees.[50]

Fabian Influence on LCC Water Policy and Its Consequences

Through the Progressives, Sidney Webb and the Fabians influenced the LCC's policy even before he was elected to serve on the county council. In 1891, Webb wrote to his fiancée Beatrice Potter that Radical ally Benjamin Costelloe had already been moving the LCC in a Fabian direction: "Costelloe . . . has prepared a tremendous draft of the Liberal programme for L.C.C. election, which is from beginning to end a 'Fabian' manifesto! . . . It goes for 'municipalisation' of every public service. . . . Liberal leaders are going to back the Progressive Party's campaign . . . and they will be committed to a most tremendous scheme of Municipal Socialism."[51] When Webb campaigned for a seat on the LCC for Deptford, his campaign fliers read, "I am in favour of replacing private by Democratic public ownership and management, as soon and as far as safely possible. It is especially urgent to secure public control of the water supply, the tramways and the docks."[52] Once elected to the LCC in 1892, Webb bore the new theory of public water directly to the committee chambers, where he was an assiduous attendee.[53] Webb very early won the respect of important councilors such as Sir T. H. Farrer.[54] As a member of the LCC's water committee, Webb helped establish council policy by voting on various initiatives, sending committee reports to the rest of the council, and so on. His influence was magnified because he served on the committee with Costelloe, his strong political ally. The Liberal and Radical Union provided Webb a venue through which to publish and speak about water for an audience that was sympathetic to his vision of an overhauled London.[55] And Webb's LCC election victory and committee appointment also opened the door to less formal influence on water matters through his socializing with the metropolis's, and the kingdom's, elite. For example, he was known to make dinner conversation that consisted of advising influential figures to take aggressive action to secure London's water supply.[56]

From the early 1890s, the Progressives' LCC adopted plans to govern gas service, water service, and transportation. The council sought

FIGURE 4. Progress through municipal works. Title page from *New London: Her Parliament and Its Work* (London: Daily Chronicle, 1895), a work reprinted from a series that originally appeared in the *Daily Chronicle*.

to direct a vast labor force, operate the port, and maintain the Thames. The council began submitting bills to Parliament for water company municipalization in the mid-1890s.[57] Income from water service, other utilities, and municipal industry would flow to the center (the municipal governing authority) and then back out again on the basis of upgrading living conditions and improving the efficient and productive operation of the city. The council also hoped to bring more sectors of the municipality under communal control. For the leaders of the Fabians and Progressives, ordering the environment was the key to social change, just as it had been for leaders in provincial cities in the past. However, the London activists were driven by a much more fundamental and ambitious vision than that of urban reformers who had gone before them. Water, retheorized by the Fabians and integrated into the wider program, began to stand for something drastically different—not social amelioration but fundamental transformation.[58] (The LCC campaign to make municipalizing water service a cornerstone of this program is the subject of chapter 4.)

In the eyes of those watching from the center of the national government—the leaders of the Conservative Salisbury government—the long-standing push for water municipalization looked radically different from this point onward. These Conservatives very quickly recognized that, in the hands of the Progressives and their Fabian allies, water reform had come to represent a dangerous slide toward collectivization. They quickly took up arms for a rival vision. Parliament had consistently granted town councils' requests to municipalize their water companies in the past, and the central government had stayed out of the issue. Now, thanks to the Fabians' communitarian vision for water, it intervened, as described in chapter 5.

Throughout the nineteenth century, scores of cities across Britain seized their water supplies with parliamentary sanction. After sanctioning the cities' takeover of water service, Parliament took rather little notice of the reforms initiated at the urban level. As long as town councils amply compensated disowned water executives and shareholders, Parliament granted the cities' private acts perfunctorily.[59] Behind this water reform movement was anxiety about preserving and promoting the liberal Victorian urban regime: reformers saw the threat of epidemic, were concerned about maintaining social order and the urban industrial status quo, and were guided by philanthropic concerns for the alleviation of suffering and the coherence of working-class families. When water reform finally came to London, the motivations were different. The Fabian Society and its concept of water supply had permeated the water reform movement.

The Fabians aimed to guide society using, in water, the same *instrument* as Liberal regimes, but, unlike Liberalism, the more illiberal Fabian socialism sought far more direct and fundamental interventions in society.[60] It was not enough to realize mere gas and water socialism—that had been accomplished in Birmingham and elsewhere without changing the underlying economic structures that compelled inequality. The material effects of water reform had to be integrated into the larger program of realigning the flow of wealth—especially the value of the land—changing the organization of labor and the provision of housing, transportation, and other resources. And needed at the center, efficiently coordinating it all, were suitable leaders—Fabian experts or acolytes, guiding society toward a new height in civilization. "Look for a moment at this London of ours as if from a balloon," Sidney Webb challenged his audience in a speech in 1892: "It is a huge manufactory of human character, a colossal breeding ground of human wills and intellects. . . . These [atoms] within the teeming life of our great city do little to make their own surroundings; they are in the main what their environment has made them. We know not who was responsible for the original creation of London. But we know that it is ourselves who are creating the London of the future. What manner of citizens do we wish this London to turn out?"[61] For Webb, the urban civilization of the future should manufacture a socialist citizenry. London should mold its citizenry's character, will, and intellect to make it community oriented, egalitarian, and enlightened. How, asked Webb, could London be governed, managed, even physically reshaped, to generate this new society? How could London be transformed into an Ideal City for manufacturing a community unlike any the world had ever known—a great commune? Water provided the key and the fuel, with the Fabians as the drivers.

The Fabian model of water supply, as one part of an integral system of public health, social justice, and socialist local government, is, as a social vision, nearly unique. However, as a plan of environmental control for the sake of social transformation, the Fabians' vision fits within a much wider context of modern water control history. Like most examples of such history, this one is about social power, if of a different current. In a trailblazing book, *Agrarian Conditions in Northern India*, Elizabeth Whitcombe describes how British colonial authorities sought to shape the Indian economy and society through canal irrigation projects in northern India. By diverting water from the Ganges over more than a million dry acres of what is today Uttar Pradesh, the British sought to change what Indians cultivated and, ultimately, to change their economic way of life. More recently, Christopher V. Hill has extended this research to British management of the river Kosi in northeastern India; there, too, colonial authorities tried to harness water to force a change in the economic activities and, ultimately, the daily lives of local farmers and their relationship with the East India Company. In these and other works on imperial water control, the use of water as an instrument of social manipulation is much more visible than in the case of the Fabian vision of water and society.[62] The mechanism is apparent because those water systems were literally on the surface. The Fabian system for guiding water and social evolution at the same time was to function mainly under the streets of London. The power of those subterranean works—works that appeared to direct no one's activities—was no less potent than that of those that transformed the surface of India.

CHAPTER 4

From Engineering Modernization
to Engineering Collectivization

N O SOCIETY EXISTS outside of nature, and nature is rarely beyond social influence. This relationship is most prominent in
the city. In the city, humans rearrange nature—the face of the
landscape, bodies of water, vegetation—for the purpose of facilitating
productivity and supporting a large collection of humans. Urban inhabitants tend to think that in cities they are more free from the forces
of nature than are their counterparts in the countryside, but that is
largely illusory thinking. It is more the case that their exposure to the
moods of nature is hidden behind a veil of the managed environment.[1]
Cities participate in nature—they are never outside of ecosystems—but
they exist within special ecosystems that have been called "urban ecosystems."[2] These systems encompass organisms, geology, and climate—
such as a wetland or desert ecosystem, for example—but also include
the human-made infrastructure and political, economic, and ideological factors found in cities. All of these factors are closely related, with
change in one factor driving change in others; nature, society, history
all feed back on one another.

Those who advocated water reform in British towns from the 1840s
onward knew that they could manipulate water to achieve particular
outcomes in society. In other words, improvement-minded leaders

could make a city "modern" (as they understood the notion) by managing the flow of water through this urban ecosystem. In their view, the city must be suffused with clean water: it must have public baths and fountains, its fire hydrants always under pressure, its streets frequently watered to keep dust under control, and all residences—regardless of status—equipped with taps. Water was also to be continuously cycled out of that environment, with water closets continually flushed and drains frequently rinsed. That water-infused environment would maintain public health, protect property, help industry, and overcome population limits. Patricians of Birmingham, Bradford, Liverpool, and other towns saw in their great environmental reengineering projects the promise of abundant water supplies through the magnitude of the waterworks, improved purity through their position in the landscape, and high pressure through the projects' design. These civic leaders saw how waterworks could be mechanisms for adjusting the urban ecosystem.

The London County Council's Progressives saw their own plan for modern waterworks the same way. In 1895, the council's engineer, Alexander Binnie, was finalizing a design for a water supply scheme like those John Frederic Bateman had overseen, only on a scale corresponding to London's vastness. He had identified seven central Welsh valleys that he proposed to flood with river, stream, and rainwater, creating reservoirs. (Once the water of approximately 500 square miles flowed into these reservoirs, they would comprise Wales's largest lakes, the largest one among the largest lakes in Great Britain.) Great twin aqueducts would convey the gathered water more than 150 miles across England to holding reservoirs above London.[3] The water would then flow down into London's water mains under high pressure. As a result, the Thames could be completely abandoned as a source of water. The plan dwarfed any other water supply project yet attempted and was among the largest environmental reengineering schemes of any kind hitherto contemplated in Britain. The LCC's water committee believed that the plan promised superabundance, a new standard of water purity, and a constant, high-pressure supply. The monumental Welsh scheme, in other words, would revolutionize London's environment.

The London County Council presented a bill to Parliament seeking the authority to execute the scheme in 1898. However, like the Metropolitan Board of Works bills for a new water source and for the takeover of London's water companies, the LCC's request incurred opposition in Parliament. The LCC, in fact, experienced far more strident and deeply rooted opposition than the MBW ever had. The county council had been a relatively ambitious body from its first day. Regardless of their political leanings, most councilors wanted London to catch up quickly with the provincial cities that had already made significant improvements in municipal services, infrastructure, and so on. Within a few years, though, the platform of the Fabians grew into the solidified agenda of the council's majority. At a steady pace, the LCC pursued powers, enacted policies, hired personnel, and created bureaucracies in an effort to realize the goals of labor reform, the reduction of the power of the rentier class, and the transformation of the urban environment. Monopolies, certain private enterprises, even some kinds of private property became targets of the council. Vocal members of the LCC's majority coalition made no secret of their ambitions to transform London from a city seemingly passed over by developments common throughout the country into a vanguard of a new governmental reform movement. Their rhetoric made the reformers of Birmingham and Bradford sound apathetic and backward. And, while Parliament had been receptive to the reformers of Birmingham when they had sought power to municipalize their water supplies, it viewed the LCC's efforts to do the same as a component of a vision that went beyond any that Birmingham had expressed. The Fabians and their allies had indeed imbued water with greater importance than ever, and a wary Parliament recognized just that.

Though a House of Commons committee had endorsed the LCC as the inheritors of London's water supply in 1891, by the second half of the decade, Parliament was unwilling to see the water supply in the county council's hands.[4] Up to this point, the government had always approved of town councils' manipulation of urban ecosystems to realize certain social goals, but now it drew the line. The central government considered the LCC's water modernization plan to be synonymous with collectivization. So, when the LCC presented a bill for authorization to build its massive waterworks, Parliament blocked it. The Welsh scheme would have placed an immense and superior supply of water—and authority—in the LCC's hands, and it would force the water companies to come to terms with the council or be replaced.

Instead of the Welsh scheme, Parliament favored a rival proposal that constituted a contrary approach, a plan for drawing water from the Thames during wet periods and holding it in reserve in vast reservoirs just west of the city. The Staines scheme, as it was called, was the opposite of the prevailing Bateman model to which modernizing towns had adhered for many decades. The Staines scheme did not abandon a river in proximity to the city, it did not introduce high water pressure through gravitation, and it did not produce a supply of increased purity. It was not, in short, a monumental work signaling a revolution in the city's environment or the improving spirit of the city's leaders. Indeed, the appeal of the Staines scheme in the eyes of Parliament's Conservative majority was that it could be built piecemeal, at less expense than the Welsh scheme, by the existing commercial water suppliers; thus, it did not demand, as did the Welsh scheme, the municipalization of the water companies.

In the second half of the decade, the anomalous behavior of the region's weather set in motion a chain reaction through the urban ecosystem. Record high temperatures and a dearth of rain lowered water levels in the companies' reservoirs. Over a series of summers, water supplies were shut down for long periods of the day in southern parts of the city and especially in eastern sections. The technical system on which water companies and consumers relied was thus affected by an unpredictable natural system.[5] This triggered social turmoil. There were public demonstrations, outraged editorials, and calls from London's second-tier local authorities demanding that London finally reform its water supply system. The effects radiated to the political sphere. By the end of the 1890s, a Conservative government that had successfully kept London's water supply from the hands of its political rivals was coming under significant pressure to reform London's water supply—to redesign London's urban ecology or stop blocking others' efforts to do so.

The Ambition and Evolution of the LCC

The first London County Council molded its aims around the view that, in the absence of an authoritative local government, London had been left prey to interests with no concern for the well-being of working-class Londoners. These interests, centered in the City or jealously guarding their privileges in the vestries, had profited from leaving London's environment unreformed. Landlords had been free to rent unsanitary quarters to poor townspeople—leaving them unconnected to gas and water services, as well—and had been free to leave their neglected properties to rot rather than paying the expense of clearing them, for example. London had paid a price, too, since without a central government that managed the metropolis as a single entity, improvements in parks, streets, and so on, had been unevenly distributed. The wealthier West End tended to receive assets while the poorer East End did not. And, of course, monopolists like the water companies had been able to operate unchecked. To address this long-standing debt to London's poor, the LCC's first informal platform called for unifying London under its leadership. It would then have the power to improve the urban environment by enacting needed reforms, redistributing resources from wealthier London neighborhoods to poorer ones, forcing action by recalcitrant landlords, and unseating gas and especially water monopolists.[6]

The first county council's program was hardly revolutionary, though perhaps the urgency with which it acted was extreme. It merely shared the vigor and objectives that reformers like Joseph Chamberlain had displayed when he was mayor of Birmingham in the mid-1870s. Chamberlain was a minister of the "civic gospel," a credo that called for the expansion of the duties, power, and wealth of local government for the purpose of addressing the inequalities that, Chamberlain wrote, "form so great a blot on our social system."[7] Chamberlain shepherded the municipalization of the city's gas supply, pushed massive slum clearances, and, of course, municipalized Birmingham's water supply—and the income from gas sales was to fund further reform projects, while income from water sales reduced rates.[8] The first LCC perceived how far behind London stood in comparison to this standard and embarked on a course to follow Chamberlain's model.

The first term of the county council, 1889–91, was marked more by its ambition than its legislative successes. The council's majority announced the aims of taking over London's water supply, of transforming working-class housing, and so on, but first the council had to expand its responsibilities and revenue through general powers bills and money bills.[9] Water supply offers one example of the council's efforts to get legislation through Parliament: the LCC first had to get a parliamentary act for the funds and authority to research water issues and to enter into official negotiations with the water companies. It also had to seek the authority to promote housing schemes, which it ultimately did in 1890, whereupon it immediately began to plan numerous projects.[10] It took quick steps to lay the groundwork for eventually municipalizing some or all of the city's tram operators.[11] It also established a policy of paying all council employees a "moral minimum" wage.[12] Whereas it succeeded in taking those small steps, other strategies simply failed. To increase income to fund a variety of reforms, the first council sought to change the way properties were assessed for tax purposes, for example, but did not succeed.[13] The LCC also failed in its effort to become the sole public health authority for the metropolis, eking out only a few additional powers while sharing authority with the vestries.[14] So ended the LCC's first term, with the body achieving little in its first three years relative to the accomplishments of Joseph Chamberlain.

The election of the second London County Council brought changes that made Parliament more suspicious than ever of the body it had created. First, the LCC's majority coalition of Liberals and Radicals, called the Progressives, campaigned on a more clearly defined, bold, and shared platform based on Fabian Society member Sidney Webb's *The London Programme*.[15] To promote their policies, the Progressives formed a body called the London Reform Union, which, for the second LCC election, issued a series of pamphlets that summarized the main points of Webb's program.[16] Fabians, following their strategy of permeation, held key positions in the group.[17] *The London Programme* offered less of a moral-political sermon based on the "civic gospel" than a vision of community approaching the status of commune. The Progressives campaigned on a policy of, among other goals, municipalizing London's dockyards and directly employing dock laborers, operating all of London's horse-drawn trams, taxing at exceptionally high rates those receiving ground rents, improving working-class housing, and taking over partial or complete control of the river Thames.[18] They predicted with confidence a victory in their efforts to municipalize London's water supply, while at the same time calling for a project for a new water supply.[19]

The reformers of Birmingham and other provincial centers had sought to improve the urban environment to improve the health, productivity, and character of townspeople, while the Progressives sought even more fundamental changes in the relationship between citizens and local government. In this new relationship, the city would be less of a mere facilitator of well-being and more the entity responsible for supplying citizens' daily needs—and being the progenitor of a socialist outlook.

London's electors seemed to embrace the growing boldness of

the Progressives' vision. Receiving approximately thirty-five thousand more votes in 1892 than in 1889, the coalition increased its numbers on the council from seventy-one in 1889 to eighty-four in 1892, which gave them a fifty-seat advantage.[20] And the composition of the new majority changed, reflecting many electors' support for the assertive program. Six members of the Fabian Society won seats on the second county council, another was made an alderman, and one socialist councilor became a Fabian during the course of the term.[21] Within the council, leadership passed from staid Liberals to more radical councilors. Leaders like the Earl of Rosebery, who left the council to serve as foreign secretary, Sir Thomas Farrer, who retired from public service, and Sir John Lubbock, who withdrew from the Progressive leadership, were the types of politicians who would have looked to Joseph Chamberlain as the model of the urban reformer. Those who rose to prominence in the second and third county councils—including Ben Costelloe, W. H. Dickinson, Charles Harrison, and Thomas MacKinnon Wood, looked more to Sidney Webb.[22] The first council's leadership had been hopeful that the body would not be strictly divided along partisan lines. Its moderate Liberal chair had called for nonpartisanship, and some principal members had campaigned and served without party affiliation.[23] That hope was not fulfilled, as the Liberals and Radicals had formed the Progressive Party during the council's first term, but the Progressive majority of the first county council placed five Conservatives in the chairs of committees and counted among the leadership conservative Liberals.[24]

With the Progressives' vast election victory, which appeared to be an affirmation of *The London Programme,* they dominated the second council and abandoned aspirations of nonpartisanship. Progressives removed most Conservatives from committee chairs, they elevated a partisan radical to the post of vice chair, and the Fabians encouraged the council to move boldly forward without regard for the opposition.[25] A number of Progressives anticipated achieving goals at the heart of their platform, including the municipalization of the water supply, within the next term or two.[26]

After the Conservatives' overwhelming defeat in the 1892 LCC election, they awoke, even at the national level, to their total loss of influence in London's first-tier government and to the zeal of those who held it all. The council had become a haunt of "Revolutionary Faddists and Spendthrifts," complained the Earl of Wemyss, a leading figure in the London Ratepayers' Defence League.[27] The council was lost to "bitter radicals," agreed a Conservative councilor in 1893, who argued that they were guided by abstract designs rather than prudence. They sought "the management of the Gas and Water Supply of London, and of the docks, to say nothing of the multitude of supplementary duties which light-hearted Fabians would confer on them . . . not for economical reasons, but in order to give effect to certain political dreams."[28] In 1894, the LCC's Conservatives, calling themselves "Moderates," assembled their own promotional organization with the aid of a number of Conservative members of the national government, including Joseph Chamberlain, now a Unionist. The London Municipal Society was to arrange public meetings, promote and elect candidates, and publish literature. It offered a positive policy of basic philanthropy, such as improvements to working conditions and housing, but at its heart was a program of opposing the growth of the LCC's power. It called for opposing the centralization of powers in the hands of the LCC and for the policing of its spending.[29] The council, in fact, increased taxes by 18 percent in its first term and was to increase them by a total of 46 percent over its first two terms.[30]

Speaking before a meeting of Conservative vestrymen and metropolitan MPs later in 1894, the Marquis of Salisbury went beyond decrying mere tax increases to call the LCC "the fortress of a political movement." He said the council was "the place where collectivist experiments are tried. It is the place where a new revolutionary spirit finds its instruments and collects its arms."[31] He offered an official endorsement of the London Municipal Society and their campaign against the Progressive LCC; the national Conservative party had officially declared war.[32]

In the same speech, the Marquis of Salisbury—who was on the verge of reclaiming his position as prime minister—praised the governments of Liverpool, Birmingham, and Manchester for their many municipal successes. The London County Council was essentially different: whereas municipalization at the hands of Manchester was acceptable to Conservatives, it was not acceptable in London with the LCC as the agency in charge. Although Parliament had seen no danger in authorizing Birmingham's town council to take local water supply into its own hands, the MP and LCC member C. A. Whitmore wrote that "it may be generally deemed inexpedient for some time to come, and until the Council has purged itself of its political distempers, to give it the huge additional labour of controlling the Gas or Water Supply of London."[33] Through the rhetoric of the growing opposition to the LCC located at the national level, water modernization was becoming linked to dangerous collectivization.

The LCC's Efforts to Purchase London's Water Companies

As part of its early efforts to purchase London's eight water companies, the LCC had asked in late 1891 for a royal commission to investigate the companies' future capabilities. The council's goal was to expose the companies' sources as deficient. The LCC's water committee suspected that the companies' main source of water, the Thames, was likely to prove insufficient or too impure to provide for London's ever-growing population, and the purchase price of the companies should reflect that liability—it should reflect the fact that the LCC might have to find a new water source within a short period of time and at great expense. In the eyes of the water committee, the companies' stock value reflected only their current capabilities and the worth of their local monopolies; the committee wanted to ensure that the companies' stock value alone did not form the basis of the buyout, as had been the case in many past instances of municipalization.

Put in another way, the council asked the Royal Commission on Metropolitan Water Supply to put the Thames on trial once again. The country's leading waterworks engineer had recommended in 1865 that London abandon the river as a drinking water source; the Royal Commission on the Pollution of Rivers in 1874 made the same recommendation, as had the Metropolitan Board of Works in 1877.[34] The 1869 Royal Commission on Water Supply had ruled that the Thames was a suitable water source, but the LCC's engineer wrote in 1890 that standards of water quality had risen since 1869 and that the science of chemistry offered better means of water analysis in the 1890s than it had a generation before.[35] The number of people living in the Thames valley had also risen since 1869, and thus it was "inadvisable to depend entirely on a supply derived from a thickly populated area, which under certain conditions, may possibly become a danger to the health of the community."[36] The county council had grounds to be confident in a guilty verdict for the river.

The council's hopes for an unambiguous condemnation of the Thames went unfulfilled, however. The Royal Commission on Metropolitan Water Supply, less formally called the Balfour Commission after its chair, Lord Balfour of Burleigh, a Conservative Scottish peer and a former head of the Board of Trade, endorsed the river as a source of drinking water.[37] After interviewing the engineers of the water companies, officials from the River Thames and Lea Conservancy Boards, rainfall experts, and geologists, the Balfour Commission concluded that the existing sources for London's water, the Thames especially, could provide much more water in the future given proper storage—up to 420 million gallons, or 35 gallons per head for 12 million future Londoners.[38] The commissioners received testimony from chemists and waterworks engineers who also attested to the water supplies' quality and the reliability of existing filtration systems. The commissioners were convinced. They were charged with exploring the availability of suitable watersheds beyond the London area if the Thames were to prove insufficient, but the commissioners judged that it was not necessary. The case, in their eyes, was closed.

The LCC's water committee rejected the Balfour Commission's conclusions, arguing that the LCC had not had the opportunity to cross-examine the expert witnesses, and so it continued its efforts to take over the water companies.[39] However, the water committee adopted a new tactic. Buy one or two of the companies, it recommended to the full county council, starting with the least wealthy, then improve their services to demonstrate that the LCC could succeed at providing a water supply just as had dozens of local governments. The purchase terms would serve as the model for future negotiations, thus easing the way for the eventual municipalization of the entire water supply. The committee recommended that the council first buy the Lambeth and the Southwark and Vauxhall Companies. They were among the least prosperous, but they also served adjacent areas of London south of the Thames. The LCC could link their source and supply networks to make their operations more efficient. With service improved, redundant directors eliminated, and other redundancies eliminated, the LCC could operate the enterprises at a lower cost. The council would increase the county taxes of Londoners to whom it provided water. If it operated the waterworks at a "profit," the council could reduce its debt or service additional loans for future company takeovers.[40] The LCC composed a parliamentary bill for authorization to purchase the companies on terms agreed to between them and the council or by arbitration that was not to be based merely on the companies' Stock Exchange value. Conservatives saw only doctrinaire pretexts for the "collectivist experiment," but with a Liberal parliamentary majority and with Lord Rose-bery, a former LCC chair, as prime minister, the LCC's bill got a second reading and a House of Commons committee hearing.[41]

But Lord Rosebery's government collapsed in the summer of 1895, erasing the bill's progress. Rosebery had been unable to advance his agenda in the face of Conservative opposition in the Commons and House of Lords, and it had enjoyed only weak support from a Liberal Party leery of his imperialist tendencies. The 1895 elections swept the Conservatives into power, and the Marquis of Salisbury returned to the prime ministry. On the eve of regaining power, Salisbury had criticized the council for its tendency to "strike out on some brand-new theory or to devise some plan unknown to the ages for relieving the miseries of mankind."[42] Once restored, Salisbury began a campaign for the par-tial devolution of the council by the creation of metropolitan borough councils that would amalgamate the old vestries. These would offer a more substantial second-tier authority that could absorb some of the LCC's duties. That fight would stretch on until 1899, when the twenty-eight borough councils were formed and began to serve as a sort of second front in the central government's battle against the LCC, which raged alongside the conflict over control of the water supply.

Climate Vagaries and the Vulnerability of the Urban Ecosystem

In late summer 1895, the LCC faced the collapse of its bills and the an-imosity of the new head of the government, but not all looked bleak for the council—the environment itself intervened to breathe life into the LCC's water campaign. Long before the LCC came on the scene, the reformers of Britain's provincial cities had argued that the modern city should expect a water supply that was constant under all conditions. And that new model city, too, demanded a source of water supply of immense volume to liberate a community from limits formerly imposed by the confines of an area's natural resources. These arguments came again to the fore when a series of water shortages struck London in the mid-1890s. Several companies found it nearly impossible to refill their reservoirs from a diminished Thames during several extremely dry and hot summers from 1895 to 1899. The LCC's water committee argued, like many water reformers before them, that water modern-ization through municipalization and waterworks construction would free London from the scarcity it was experiencing. The water fam-ines appeared to argue against the Balfour Commission's picture of the

Thames and the Conservatives' antagonism toward the county council.

Responding to calls for help on a balmy day in early September 1895, the fire brigade tapped a nearby hydrant to fill its tank. Finding the hydrant dry, they moved down the street to tap the next, then the next, and next. Finally, at a distance of 165 feet, the firefighters discovered a flowing hydrant.[43] A "special correspondent" covering London's new water crisis for the *Times* commented, "We are confronted by scarcity due to hot weather, and the water companies are once again the common cockshy for indignant consumers."[44] The metropolis had suffered a shortage of rainfall for many months. During the summer, the area's rainfall was running 5.83 inches below normal.[45] The large deficit remained into fall, with little sign of London's usually revivifying autumn rains; by the end of September, the region had only drawn closer to the average by an inch.[46] High temperatures that September were frequently above 80 degrees and rarely below 74. Some of the peak readings included temperatures that were the highest hitherto recorded at such a late date in the season.[47]

While people marveled at the unusual string of cloudless, hot days in June and July, the water level in the East London Water Company's reservoirs began to drop visibly.[48] To conserve supplies, the company stopped pumping water during the nighttime hours. As July wore on with no respite in the heat or drought, service hours were drastically cut, from fifteen hours to three.[49] Customers reported receiving water only from nine in the morning until noon.[50] Relief was slow in coming. It was not until the first of August that service was increased to six hours and only after mid-August that it was increased to eleven hours a day. Customers were affected for a total of two months and nine days, and all the while they had to pay the same water fees.[51]

The water companies, it bears repeating, were not statutory monopolies under any Parliament-imposed obligation to provide water to Londoners, nor were they under any contractual obligation to their customers to supply them with water. Customers either paid their water rates or were promptly disconnected from the water main. Companies were simply private enterprises that had arranged local monopolies among themselves. Their few obligations to the government were defined piecemeal, through the occasional parliamentary acts. After passage of the Metropolis Water Act of 1852, the companies had to filter their water, draw it from places other than the tidal Thames, and get consent from the Local Government Board before they drew water from new sources. Under the Metropolis Water Act of 1871, they had to provide water of a minimum standard as determined by a government-appointed metropolitan water examiner, who, in practice, never gave the water companies much trouble.[52] In another act of negligible regulation, Parliament passed another water act in 1897, and it allowed individual customers or local authorities to bring a case before the Railway and Canal Commission in the event of a water company's negligence. In practice, few cases were brought, and when they were, it was extremely difficult to prove any company's negligence.[53]

The summers of 1895–99 were the warmest recorded in forty years and the driest in thirty years.[54] The East End's water shortage of 1895 was immediately followed by two more summers of drought and water rationing (with the 1897 crisis hitting South London especially hard, as well as the eastern part of London). These crises, though, were exceeded by the water shortage of 1898.[55] By mid-September of that year, the region was no less than seven inches below its mean annual rainfall, and springs supplying the East London Company began to dry up for the first time in living memory.[56] The company resumed its water rationing, and, toward the end of August, it posted fliers in East London neighborhoods declaring,

> In consequence of the severe and continuous drought, NOTICE IS HEREBY GIVEN that the supply of water will be restricted on and after Monday, 22nd August.
> The water will be turned on twice a day for about three hours each time, and at the same hours as nearly as possible daily.
> Consumers are advised to fill any available vessels while the water is on, to use it only for strictly domestic purposes, and to avoid WASTE in any form. Persons are especially cautioned against using water for gardens or other similar purposes.[57]

Soon thereafter, vestry water carts were again sent through the neighborhoods.[58] Some residents reported receiving no water at all, let alone a limited supply, while twenty-three schools reported problems such as drinking water restricted to two hours daily and inability to flush water closets.[59]

The problem of unflushable water closets was of particular concern to local authorities. The Hackney Vestry posted notices suggesting that its residents keep a large jar on hand to capture what water they could when their taps were flowing and also requested that residents "see the Water Closets and Sinks are well-flushed every time the water was on." That left eighteen hours a day that a water closet, usually the sole water closet for one or more families, could not be flushed. The vestry tried to offer some relief, advertising, "Disinfecting Powder for Water Closets may be had on application at the Town Hall—FREE OF CHARGE."[60] Inoperable water closets were particularly distressing to a medical community for whom the first rule of public health was to remove human waste as quickly and as far away as possible. Doctors in the East End, for example, reported an increase in cases of diarrhea that August and warned in an open letter that "the stoppage of the water supply will directly cause . . . deaths."[61]

Turning Natural Events into Political Capital

Having suffered a setback at the hands of the Balfour Commission and facing a hostile national government, the LCC's water committee sought to harness public exasperation with water shortages. Individual councilors personally brought the companies' failed handling of water shortages to the notice of consumers. For example, George Balian, an East End council representative, helped lead a public protest during the 1898 drought and moved "that this public meeting . . . is of the opinion that the time has arrived when the control of the water supply should be in the hands of the London [County Council]."[62] W. C. Steadman, a Fabian member of the LCC, was president of the Stepney Labour League, which resolved that the East London Company should "give more consideration to the health and lives of the people in the East End than to their greed for dividends."[63] Thomas Idris, a Progressive who served on the council's water committee, chaired the Christian Social Reform League, which denounced the companies' failures and called for a bill empowering the LCC to take over the water supply.[64] Speaking during the drought of 1898 before a meeting of the London Reform Union, W. H. Dickinson, chair of the LCC water committee, claimed that "what had been happening during the last weeks in the East-end [will] occur in a few years throughout London. . . . The only solution of the problem lay in the ability of the LCC to [offer a water supply]."[65] The council, Dickinson suggested, would be able to resolve the problem because it would be above the profit motive guiding the companies and could manage London's many waterworks as a single network; because an LCC-operated system could compensate an underprovided area with water from another, there was an implicit claim that the LCC would somehow be immune from the threat of shortages.

In weekly county council debate, partially transcribed in the city's larger morning dailies, the Progressives sustained this refrain. James Stuart, an East End councilor and Liberal MP, claimed that "the present distress in the East-end was due to the culpable negligence of the East London Company."[66] His fellow Progressive, George Shaw-Lefevre, declared that "the events of the past few weeks had vindicated the water policy pursued by the Council. Had the proposals of the Council been carried, there would have been no scarcity in the East of London." The Hackney representative, Alfred Smith, concluded that "there was no remedy for the present state of affairs other than the municipalization of the London water supply."[67]

Behind closed doors, the Progressives calculated the effect that the droughts would have on their political aspirations. In a confidential memo sent to the LCC's water committee, its chair, W. H. Dickinson, assessed the consequences of the cycle of droughts and water shortage of 1898 in particular. "The special circumstances of the last few months are twofold and affect the question as to what legislation is necessary in two manners," he wrote, adding, "First, there is the fact that one quarter of the population of London has recently been subjected to a series of water famines by reason of the default of the East London Water Company."[68] The seven East End parishes and district boards had a combined population of 806,102.[69] Thus, hundreds of thousands of prospective electors in the East End had been inconvenienced, potentially giving the LCC an undeniable mandate and pressuring London MPs to side against the water companies. "Secondly," Dickinson continued, "there is the fact of the drought of 1898 which has shown to what extent the supplies generally available for the metropolis can be, at times, reduced."[70] There was a degree to which the inconvenience, or at least the threat of it, was general throughout the vast city.

Dickinson believed that, in addition to earning the companies a good deal of ill will, the water shortages had made the case that the LCC was the perfect inheritor of the metropolis's water supply. During the drought of 1898, the East London Company had been criticized for failing to obtain a supplemental supply from any of the other metropolitan companies.[71] While the company had publicly responded that such a plan was impracticable from a technical standpoint, it made private overtures to other companies nonetheless. Its appeals were denied. Water suppliers such as the New River Company needed to preserve their own resources for their own customers or risk water failures and public animosity.[72]

Dickinson appeared to have known these circumstances or anticipated them, writing, "If the Council's bills had become law in 1895 there would have been no water famine in the present year [while] . . . the continuance of the individual companies is open to the objection that each company, possessing a margin of available water, will naturally desire to retain such margin for the possible needs of its own consumers."[73] If the LCC owned all eight companies' works, the council could interlink the systems of distribution and would not be deterred from doing so by the profit motive.[74] It was a propitious time to bring forward new bills for the acquisition of the companies, Dickinson concluded.[75]

The water companies watched their political vulnerability rise as the levels in their reservoirs dropped. They argued that they could not possibly have made accommodations for such an exceptional drought. They pointed to historical rainfall data and the opinions of experts who attested to the exceptionality of the dearth of rain. It was simply a fact of nature, rendering "eminently absurd . . . the agitation being raised." The meetings of indignant consumers were mere "concoctions" of the county council, staged to influence "the politics of the drought," which could not, after all, make it rain.[76]

Despite the water companies' protests, the extreme weather, combined with the Progressives' representation of the crisis, motivated support for water reform. Public demonstrations called for the elimination of the companies and often called for placing the water supply in the hands of a representative body. The Progressives had sponsored or permeated many demonstrations, but there appeared to be just as many that were unprompted by the LCC.[77] After the drought of 1895, a group met at the Camberwell public baths in South London and resolved that "the health, convenience, financial interests and general well-being of the Metropolis demand the immediate transfer" of the water supply to the LCC.[78] In September 1898, a town hall meeting of the inhabitants of Leyton in East London demanded that the government place London's water supply in a "popularly elected public authority." On about the same date, a public meeting held in an East Ham park called for the dissolution of the companies as well. During the 1898 drought, the

East London Water Consumers Association formed "for the purpose of trying to get [the] government to pass a Bill so that the County Council or a municipal body may have powers for the control of the water supply." The group sent a delegation to the president of the Local Government Board to voice its frustration during the worst of the 1898 water shortage.[79] The Ratcliffe Vestry called for public control of the water supply, the St. Mary, Stratford Bow Vestry designated the LCC as the appropriate future water authority, and the vestry of Mile End New Town called for the LCC to take over the companies and secure a new source of water supply.[80]

The vestry resolutions condemning the companies were especially promising for the LCC. What the LCC needed if it hoped to win an act of Parliament allowing it to seize the water companies were the votes of London's electors, who would not only keep the Progressives in control of the LCC but would also dictate to London MPs what it needed from Parliament. With each passing water failure, the LCC did indeed count more and more vestries as like-minded enemies of the companies, even though they tended to be conservative in stance as a whole. During the drought in the summer of 1895, the district boards of Limehouse and Whitechapel called for the LCC to replace the companies in controlling water service.[81] In the East End, the Hackney Vestry called the 1898 water shortage a "dire calamity," while the parish of St. Mary, Stratford Bow, condemned the failures of the East London Company and declared that placing the supply in the hands of the LCC would be "the only effective remedy."[82] There were many such resolutions passed, and their number peaked during the worst shortage of 1898.[83] The results of the 1898 county council election, in which the Progressives recovered from the mediocre performance of 1895 to restore its large majority, suggest the popularity of the party's policy, including its campaign against the companies.[84] Meanwhile, the leader of LCC conservatives said that former antimunicipalization "stalwarts" were beginning to bend to apparent public will, "legislating," he complained, "in a panic."[85]

Another way to understand the means by which natural systems affect the social system—in this case, how the extreme weather elicited support for water reform in London society—is through the model of the urban ecosystem. Londoners received their drinking water through technical systems—the apparatus and administration of the water companies—and through natural systems—rain and rivers—that joined together in London's urban ecosystem. Every time they turned on their taps, Londoners were participating in this intricate system. Components and participants like population, rain, river flows, and reservoirs all had links to politics, economics, and ideals. Sometimes, this close interaction made components throughout the web especially vulnerable to disaster.[86]

Beginning in the mid-1890s, an unusual series of events—the extended drought and heat—combined with the other elements in the web to create a crisis. Angry water consumers wanted to lay blame for their hardship on someone. They looked at their dry taps and made the connection between dry taps and the water companies. Townspeople did not censure cloudless skies; they berated the companies and their apparatus, which many Londoners judged did not make provision for natural unpredictability. So the population caused the repercussions to be experienced by other members of the urban ecosystem. The anger of individuals, vestries, and consumer organizations fed back through the system to water companies and to the political entities and arrangements that supported a status quo that had done nothing to prepare for a catastrophe waiting to happen. The LCC, by encouraging protests, adding condemnation of their own, and offering solutions that served their vision for London, sought to manipulate this feedback and take advantage of it.

The Welsh Scheme

The Progressive-dominated LCC made great use of these recurrent water shortages in its efforts to win support for new waterworks. In the past, water reformers had usually coupled calls for water supply municipalization with demands for new water sources and importation works; the LCC, true to its character, quickly developed ambitions based on an even grander vision for the modern city. From its earliest days, it began slow and steady inquiries into a water supply scheme. The plan at which the council arrived quite simply put the monumental works of Glasgow, Manchester, and the like to shame. It was a scheme that stretched farther into the hinterland, harnessed a larger watershed, and promised a greater volume of water than any scheme yet designed. The water companies' failures in the middle of the decade provided the LCC a powerful justification for bringing forward such a colossal plan, and the central government was forced by the supply failures to give the council's proposal a hearing before initiating yet another royal commission on London's water supply. Recognizing, though, that the project of water modernization would serve the council's goals of collectivization, the central government endorsed a rival engineering scheme propounded by the water companies, one that promised, instead, to preserve the status quo.

When the council first created the forebear of its water committee, it charged that group with the dual task of inquiring into whether to acquire the companies, how to do it, and whether the council's engineer should investigate new sources.[87] Given the legal limits on the council's activities, however, its new engineer had a limited range of options to explore. He investigated what he could without spending extra funds, gathering modest amounts of information about how other towns secured external water supplies and so on.[88] When Parliament granted the council the authority to spend funds on its water supply investigations in 1890, the chief engineer immediately appealed to the council for permission to send investigators to Wales to make detailed studies of the possibilities for a Welsh supply.[89] It is not surprising that Wales was the immediate target of a study; the council's engineer, Alexander Binnie, was with John Frederic Bateman as a pupil on his surveys of northern Wales in the early 1860s.[90] Binnie also had had experience as a railway engineer in southern Wales in his early twenties.[91] He recorded that "as far back as 1862 I had observed the capabilities of the districts of the [rivers] Wye, the Ithon, the Usk, and the Towy for a large supply of water to the metropolis."[92] Binnie reported to the council that he had analyzed the suitability of his teacher's 1869 Welsh scheme, assessed and considered the 1869 royal commission's objections to the original project, and judged them easily overcome.[93] He wrote that he still had every confidence in the effectiveness and practicability of Bateman's scheme, plus he had the added advantage of time. "So long a time has elapsed since 1869, so many alterations and improvements have been made, and so much new knowledge [has been] acquired," he wrote.[94] The council agreed, and Binnie immediately sent a small team of researchers to Wales to conduct a study of rainfall amounts to make detailed maps of watersheds.[95]

One reason Binnie focused on a Welsh source was the presence of competition for Welsh watersheds. If London did not act to purchase land for reservoirs and aqueducts soon, it might find itself obstructed by a rival. Already in 1880 a legal authority on water supply matters had written that "a glance at a 'large map' of England will show that very few of the valleys of the 'backbone of the country' . . . are unappropriated by some town or company, and unoccupied by some large reservoir."[96] Since Bateman promoted his original Welsh scheme, for example, Liverpool had purchased some of the watersheds he had advised tapping.[97] And while Binnie was putting together his own Welsh scheme, he learned that Birmingham was also reaching for the Welsh hills. "I should inform the [LCC water] committee," he warned them in a letter, "that it has come to my knowledge that the Corporation of Birmingham have instructed their Engineer to [submit] to them a scheme for supplying their town with water from the district of the Upper Wye."[98] He continued, "If the Council do not succeed before long in obtaining a footing on the ground[,] in a few years it will become impossible for London to obtain an unimpeachable gathering ground."[99]

There was a veritable scramble for watersheds, and the LCC took its engineer's warnings to heart.[100] From early on, the water committee advised the council to move quickly forward with its research into the Welsh water supply because of the threat of being cut out of the rush for Welsh watersheds. "To leave Birmingham unopposed," the committee cautioned in a report to the rest of the council, "may be to lose the chance of an excellent source of supply for London and . . . nothing can be more unsatisfactory than to leave the appropriation of the few remaining areas in England which are capable of affording a good water supply to a scramble between different local authorities."[101]

As the LCC feared, in the spring of 1892, Birmingham did bring a bill to Parliament for authority to purchase land in mid-northern Wales for a new source of water. The council at that point was still in the early stages of devising its plan. Its only grounds for opposing Birmingham were that it wanted to keep Wales available as an option. The water committee hosted more than twenty London MPs to strategize about how to deal with Birmingham's challenge.[102] While London was not in a strong position, the group saw an opportunity to, in a manner of speaking, divide the water of Wales between London and its chief rival. London's MPs, including the London County Council members George Shaw-Lefevre, James Stuart, and Sir John Lubbock, then called a confidential meeting with Birmingham's MPs. London, they suggested, would not oppose Birmingham's plans to purchase land in north-central Wales as long as Birmingham would not oppose London's eventual plans to purchase watersheds in the south-central part. They also proposed that Birmingham, as it planned its future waterworks system, should cooperate with London, consulting London before tapping streams the LCC might need and avoiding interference with the LCC's envisioned aqueduct courses. Both sides appeared satisfied, resolving "to let it be understood as a matter of Parliamentary arrangement that in consideration of the County Council using their influence to support the Second Reading of the Birmingham Bill the representatives of Birmingham should in like manner use their influence in support of the [LCC's purchase efforts.]"[103]

But London and Birmingham's entente came apart only days later, when a speech by the chair of Birmingham's water committee became public. He had recently argued that "there is a most important reason for no[t] dallying with this matter [of purchasing a watershed in Wales], and that is what is being done by London. . . . Our rival they will be sooner or later, and we have just now an opportunity of running before them, and getting hold, if we can, of these valuable rivers for the supply of Birmingham."[104] Birmingham's bill, though, passed a second reading without the help of London's MPs and despite the objection of one Welsh MP, who complained that "the Members for Birmingham and London regard Wales as a carcass which is to be divided between them according to their own wants and wishes."[105] Birmingham ultimately secured its desired act and began construction on its immense Elan valley waterworks in 1893.[106]

In addition to securing one of a dwindling number of water sources, a Welsh scheme presented an opportunity for the LCC's Progressives to gain the upper hand against the water companies in their efforts to purchase them. Simply put, if the LCC offered cleaner water, under constant pressure, and at lower rates, the water companies could hardly ask an inflated price for their suddenly redundant companies. George Shaw-Lefevre, a member of the LCC's water committee, also counted on the Welsh scheme to seize the companies' attention. He wrote that "indeed, when the Welsh scheme was first seriously propounded by the London County Council, the Companies at once perceived that it threatened their very existence."[107] Sidney Webb hardly veiled his threat to destroy the companies if they would not sell their holdings on fair terms, writing in 1891 that "London is not bound by these extravagant estimates [of the purchase price of the water companies]; and the London County Council may, if it chooses, give the companies the go-by, and imitate Manchester and Liverpool in seeking for itself an unpolluted supply from afar."[108]

Even before he was elected to the LCC, Webb called for "an aqueduct from the Welsh hills [as a way to make the] 'water lords' see their polluted supply made obsolete."[109] And, running for reelection to the LCC after three years of serving on its water committee, Webb wrote to his constituents in Deptford that

> the committee sees its labours near fruition in the presentation to Parliament, of eight Bills [in fact, two bills for the purchase of the Lambeth and the Southwark and Vauxhall Companies] for the purchase of the water companies' undertakings.... During the next few months the battle will be fought in the committee-room of the House of Commons, against all the forensic talent and expert energy which wealth can enrol in the defence of monopoly rights. But ... water companies have been beaten before, and may, in a democratic Parliament, be beaten again. The Thames is not the only, nor even the best, source of London's supply, and when the time comes the Water Committee will show that its prolonged investigations for the protection of the ratepayers have not been thrown away.[110]

Indeed, Webb almost never wrote about London's water supply without referring to a Welsh scheme. It was integral to his vision of water reform and integral to his vision of social reform. Most obviously, a municipally owned alternative water source served his goal of social reconstruction by making the companies' works obsolete, helping to get the water supply in the hands of the LCC, and thus realizing all the benefits of the municipal water supply the Fabians envisioned.

Beyond that, the Welsh scheme promised intrinsic advantages of its own, or rather, the scheme would bring new benefits when put to new uses by the Fabians' allies. Webb promoted the municipalization of London's water supply in part because of its potential to expand the municipal workforce and ultimately improve conditions for all workers. The Welsh scheme promised to expand that workforce dramatically. In 1895, Webb pointed to Birmingham's ongoing Welsh water scheme as an example. Birmingham, he wrote, "is now actually engaged in constructing several huge dams and reservoirs near Rhyader, two tunnels and various water towers and siphons, together with workmen's dwellings to accommodate a thousand people ... all without the intervention of a contractor."[111] Webb pointed out that Liverpool, too, used an army of directly employed labor for its water supply project.[112]

One of the "improving" elements that supporters claimed the

Welsh scheme shared with previous projects was its integral capacity to increase the use of water among the lower classes. As a gravitation scheme, it could supply water under pressure and thereby allow water to be delivered to every inch of the metropolis. The water companies were notorious for their unwillingness or inability to provide water to the upper floors of houses and tenements, especially those located at some height above the level of the rivers Thames and Lea. Steam-powered pumps forced water through the London companies' mains, as was the case in most British towns. But despite the insistence of engineers like Thomas Hawksley, who argued that the existing systems could be made to withstand constant pressure, the companies maintained pressure only for certain hours of the day or even days of the week.[113] In 1895, just three-fifths of London houses that had water at all received continuous service. And for those who received no service at all because they resided in upper-floor rooms, their only recourse was a pump in the street.[114] The Welsh scheme would provide constant service in addition to high-elevation service. "We see in imagination the County Council's aqueducts supplying London with pure soft water from a Welsh lake," Webb wrote, "the County Council's mains furnishing, without special charge, a constant supply up to the top of every house."[115]

Without the handicap of low water pressure, the LCC could demand that landlords provide their working-class tenants with greatly increased access to water. "[Soon] the landlord will be required," he wrote, "as a mere condition of sanitary fitness, to lay on water to every floor, if not to every tenement, and the bath will be as common an adjunct of the workman's home as it now is of the modern villa residence."[116] Along with equality in access to water, the plentiful, constant, pure supply would bring public health benefits. "A constant supply of pure soft water . . . [would allow London's] millions unlimited opportunity to wash," Webb wrote. If the LCC were to provide numerous municipal baths as an accompaniment to its new municipal water supply from Wales, Webb added, the project "could not have other than beneficial consequences on the public health."[117]

From Visions "in Imagination" to Policy

Under threat of competition from Birmingham, the London County Council ordered its engineer to continue developing a plan for Welsh waterworks. From late 1891 until 1893, Binnie kept his surveyor and rainfall analyst on the job and traveled to Wales himself on occasion to oversee their work.[118] Binnie's attention shifted away from Bateman's proposed gathering grounds in north-central Wales to watersheds in south-central Wales, but he otherwise kept close to his teacher's model. When the council's water committee ordered Binnie in February 1894 to prepare a report on the best site for drawing a water supply, along with the method of delivering it to the metropolis and estimates for the project's cost, the engineer drew heavily on Bateman's investigations.[119] By June 1894, Binnie was ready to give details. First, his engineers were to construct seven new reservoirs in the hill valleys of the Welsh interior. These would be formed by damming up narrow hill basins or by transforming existing lakes into embanked reservoirs that would hold water in reserve until it was transported to London. Some of the reserved water was to be redirected back into the streams of Wales during dry spells to maintain the water levels in the Welsh watercourses. The water that London demanded would flow in two aqueducts of unprecedented scale across England. They would be constructed of masonry and concrete for much of their length and partly of iron or steel pipes in stretches where the aqueducts spanned valleys. To avoid contamination, they were to be covered for the entire length, through use of tunnels or cut-and-cover channels. At 16 feet wide and 11 deep, they would transport 200 million gallons per day. After covering 150 miles, the water would empty into two additional reservoirs built on two hills about 300 feet above the metropolis. These were to be precipitation reservoirs, where naturally occurring debris would settle to the bottom and where the water would pass through a sand bed to be filtered for clarity. The water would then descend to London under gravitational force, capable of delivering water to the upper stories of buildings without the need for pumping. The existing water company networks were to be integrated into the new system, if possible.[120] In short, the Welsh scheme could afford the elimination of the companies since their distribution system could be appropriated while their sources were made obsolete.[121]

Binnie described the Welsh supplies as pristine, writing, "As regards quality of water, there can be no doubt that from its sparsely populated areas and steep mountain slopes a quality of water can be obtained in every way desirable for the supply of a large population."[122] Binnie explained that "the waters in question running as they do in their natural state . . . are all of greater purity than the water supplied to London after filtration, and . . . being drawn from mountain slopes they are not liable in the future to be contaminated by animal or town refuse, so that the possible danger which must always hang over a water supply derived from such rivers as the Thames." Not only were the waters of the Welsh watersheds supposedly devoid of invisible dangers but they promised to remain free from impurities. And the engineer added that storing the new supplies in new reservoirs built into the valleys of Wales would only improve their quality. "The natural purity of the waters of these districts will be much enhanced by the storage to which they will be subject in the large open reservoirs exposed to pure mountain air," he explained.[123] It was as if the wholesomeness of the distant Welsh landscape could be bottled and delivered to the city from a sort of "water preserve." Binnie was arguing from the evidence of aesthetics. In reality, it was certainly possible that water drawn from a river or wells could have been as clean as water derived from green Welsh hills, but it was as if Binnie was arguing that it could never be so *wholesome.*

He was not alone. And one did not have to be an engineer with a vested interest in the results of the argument to express the opinion. Earlier, other non-engineer observers depicted Welsh waters suitable for England's cities in virginal terms: "beautifully pure and soft," "uncontaminated, and unharmed."[124] Others suggested that Liverpool's Vyrnwy water supply, drawn from upland valleys and supposedly preserved from the filth of the city, was particularly "uncontaminated."[125] And others waxed poetic about the waters available in Scottish hills.[126] It was not that there was no argument about whether only upland waters were suitable for cities; there were many who believed that deep wells would suffice and that rivers had self-purifying abilities.[127] But, despite the continued use of local, river-drawn supplies in many towns, the belief in the superiority of hill-derived water was ascendant among non-engineers. Doctors, the Royal Society of Arts, and various royal commissions declaimed against river-derived sources.[128]

Promoting the Health and Abundance of Welsh Water

Binnie also described his Welsh scheme as the ideal answer to the previous generation's anxiety over water famines. The Welsh supply was so bountiful, in fact, that "the metropolis does not yet require so large a supply as that which can be obtained."[129] The Thames valley watershed of 3,542 square miles provided 300 million gallons a day, while the Welsh watersheds of 488 square miles could provide 415 million gallons of water a day; that would have been more than the total amount of water that flowed over the Thames' Teddington weir on the average summer day; in other words, the Welsh scheme could provide London with an entire Thames-worth of water daily.[130] The river could be completely abandoned as a water source, Binnie concluded. The water supply from Wales, plus a small supply of water from wells sunk in the chalk layer of Kent, would provide "a total possible volume of 482 million gallons a day, which at the present rate of 35 gallons per head per day would afford a supply to a population of nearly 13,800,000 persons."[131]

The obvious tactic for promoting the Welsh scheme was for the LCC's Progressives to present it as a corrective to recent water shortages. "Millions of human beings . . . have been suffering during the last few months," wrote Dickinson in a pamphlet published in late 1898, "and for this reason we have formulated these proposals."[132] He warned that "unless we start at once with our Welsh Scheme we cannot safeguard London, and under these circumstances I can hardly believe that Parliament will refuse us permission to proceed."[133] Wales, it seemed, was free of the climatic quirks that dogged London in the 1890s. Progressive-allied writers continued their support for the project as well. A month after Dickinson's pamphlet appeared, the *Daily Chronicle* printed an extended description of the plan and concluded that "two alternatives . . . lie before London. On the one hand are the companies, at the end of their resources, breaking down in winter when it freezes and in the summer when it doesn't rain. . . . On the other hand is the never-failing supply of fresh, pure, bright water, supplied at low price by the community itself."[134] In the face of recurring shortages throughout the remainder of the decade, more vestries, associations, and individuals appealed to the LCC to eliminate the companies, abandon the Thames, and procure a new water source.[135]

Another tactic was to condemn the poor quality of existing sources while extolling the purity of a potential Welsh supply. "The fact seems to be that we in London go on being content to use river water taken below vast and populous areas of sewage pollution, only because our forefathers drank something worse," complained Sidney Webb's ally Benjamin Costelloe in the pages of the *Contemporary Review*, but it is "unsafe for an enormous community to go on drinking polluted river water, as in fact every great city except London has recognised long ago."[136] The Progressive water reformers repeated the refrain that London lagged behind other cities in water affairs. W. H. Dickinson, a dogged promoter of the Welsh scheme, echoed Costelloe in the *Contemporary Review:* "when this great problem is regarded as a whole, it is cheaper to supply pure mountain water from Wales than to store and supply the present more or less contaminated Thames water."[137]

The Welsh water source promised to be not only more healthful than a Thames supply but also more cost effective. And the present water supply, Dickinson warned, depended on careful treatment in filter beds for its quality—a potentially precarious arrangement. "The safety of health in London depends entirely upon the treatment of water, and . . . if by any means this purifying treatment were to fail to take effect, or if there were such disease in the upper portions of the rivers as to make it impossible to prevent germs from passing into the consumers' houses," he wrote, "Londoners would be obliged at enormous expenditure to abandon their river supplies."[138] The Fabians, who placed so much hope on the project, also advocated the Welsh scheme in their tracts. Fabian Tract no. 81, *Municipal Water,* took up the usual refrain of pointing to the provincial cities that enjoyed imported water supplies. The young economist Charles Knowles described the affordability of Birmingham's Welsh scheme, the abundance of Glasgow's water supply thanks to its Loch Katrine project, and how Manchester had total confidence in its ninety-five-mile aqueduct.[139]

The scheme's Progressive promoters took advantage of calls for, and expectations of, a new water supply to claim that only the LCC could accomplish the undertaking.[140] Only a visionary government, thinking and working on the scale of the whole metropolis, could manage it. "The problem must be grappled with as a whole," wrote Dickinson, "and it cannot be expected that eight separate private companies will . . . take so bold a step as that which has been taken by the Corporations of Glasgow and Birmingham."[141] The companies' interests were too fragmented. He added that "municipal bodies are far more willing and able to embark on the bolder schemes that are requisite for the health of the community than are private companies. The great aqueducts that conducted the mountain waters to Rome were mostly constructed in the time of a republican government."[142] Companies concerned with maintaining profits in the short term would never undertake a project so immense as the Welsh scheme. "It is not safe, in the interest of the health and other interests of the public, to leave the water supply of the

Metropolis . . . in the hands of private companies . . . whose financial interests are the first and main consideration to them," asserted George Shaw-Lefevre, a member of the LCC's water committee. He added that "the work of carrying out a great Welsh scheme can only be carried out by the London Council."[143] The county council was in a unique position to deliver what the drought-weary public called for.

The LCC and the Welsh Scheme on Trial

In 1897, a new royal commission began to deliberate whether Parliament should grant the LCC the power to carry out its proposals. Appointing yet another royal commission to study London's water supply—the third such royal commission in thirty years—served the Conservative government's interests in several ways. Public outcry against the water companies made action, or at least the appearance of action, politically imperative.[144] LCC Conservatives and metropolitan MPs feared that public opinion had moved so strongly against the water companies that inaction invited political disaster. In a confidential memorandum circulated among Salisbury's cabinet, the chair of the Local Government Board reported that the London caucus's belief was that "London is strongly in favour of 'something being done' to take control of the Metropolitan Water Supply out of the hands of the Water Companies. . . . If 'nothing is done,' the Moderate Party and the Unionist Parliamentary Party will be seriously injured in the judgment of the London Municipal and Parliamentary constituencies."[145]

The government, too, had declared its intent to take steps to resolve the metropolitan water problem at the beginning of the 1896 session. It could, and did, use this promise of future steps to justify opposing the LCC's bills in the second half of the decade, but it could only allude to impending efforts for so long until it had to make a show of action. A royal commission by its very nature demonstrated that the government was studying the problem, that it was on the cusp of recommending to Parliament a distinct course of action. In 1897, the government calculated that it did not have sufficient support for a bill of its own, a bill that would consolidate the water companies while keeping them out of the hands of the LCC. The companies were not satisfied with the terms of purchase, and too many of the counties around London were on the side of the LCC.[146] Quite simply, a royal commission bought the Conservatives time to consolidate their position while making opposition to the LCC's action—on the grounds that the expert commission had yet to report its findings—look prudent instead of obstructionist.

The 1897 commission was commonly called the Llandaff Commission after its chair, Henry Matthews, Lord Llandaff. A Conservative appointed by Salisbury's government, the seventy-one-year-old Matthews had been home secretary from 1886 to 1892 and an MP for East Birmingham from 1892 to 1895. His commission was to take witness evidence, collect data, and consider the question of whether and how the water companies should be purchased by a public authority or authorities, or whether additional powers should be given to local authorities to oversee the water companies. The commission was to scrutinize the potential magnitude of London's future water demands, matters of distribution throughout the growing metropolis, and the financial issues involved in any potential buyout of the companies.[147] Llandaff's commission, assembled in the spring of 1897, met for two years.

The Llandaff Commission provided the LCC, whose personnel would testify before it for many weeks, an opportunity to present itself as the ideal authority to take over the companies, and the council's Welsh scheme was the bedrock of its case. It needed to persuade the commission, or at the very least the London public (which could read the thorough proceedings in the *Times* and other newspapers), that the council had an affordable, practicable scheme that would answer the complaints of the ratepayers and consumers. If the commission looked favorably on a Welsh supply, Parliament, or even a Conservative House of Commons, would appear unsympathetic toward London in rejecting the LCC's bill. Dickinson predicted the water shortages of the decade would force the Conservatives to give the LCC's proposals a fair hearing.[148]

In opposing the LCC's plan, the companies primarily put forth the suggestion that the Welsh scheme was extremely expensive. The LCC's final report on the proposed project put the total cost at £38.8 million.[149] The companies commissioned their own engineers to assess the probable cost of the scheme and arrived at estimates ranging from £42 million to more than £52 million.[150] As a point of comparison, Britain's annual expenditure on the Royal Navy was £28.5 million in 1900.[151] The companies' commissioned estimate reported not only that the project was prohibitively expensive but also that it was not feasible to calculate the scheme's exact cost with accuracy. Cross-examining Alexander Binnie before the commission, Edward Pember, the water company counsel, pressed the engineer to produce precise costs for various components of the project. Ostensibly seeking information on the price of the Welsh aqueduct per mile, Pember caught Binnie unable to produce a figure readily:

PEMBER. You must excuse me for saying that there are at least half a dozen estimates in the notes?

BINNIE. I know there are many, and it is for that reason that I find difficulty in replying to your question.

PEMBER. They are all from your side, and they all differ. There is your own and there is Mr. Deacon's?

BINNIE. Take me on my own.

PEMBER. Mind you, I might ask which of yours you mean because you had one the other day, and you got one this morning?

BINNIE. I have got to answer the questions put to me, and I answer them to the best of my ability. I know the questions are intended to confuse by being shuffled about; but if you tell me which you are cross-examining upon, I will answer.

PEMBER. Really, my Lord [Llandaff], I do not know that I am susceptible, but is the Witness justified in saying that?[152]

The exchange testifies to the heatedness of the confrontation between the LCC and the companies (as well as to Pember's noted tendency to bicker with witnesses).[153]

In addition to attacking the LCC's project on the grounds of its large or uncertain expense, the water companies and Conservative witnesses criticized the county council itself on political-ideological grounds. The companies and the Conservatives shared an interest in defeating the Welsh scheme—the water companies simply sought to survive or, failing that, to have their purchase undertaken by anyone but the LCC, which had endeavored for years to drive down the companies' price. The Conservatives, of course, simply wanted the water supply kept out of a Progressive LCC's hands. The LCC was "too political a body to have the management of the water supply," argued C. A. Whitmore, a Conservative MP and LCC member, before the commissioners.[154] A water company representative cross-examining him agreed that "a large number of what are called Progressives [have] an ideal before them of managing vast undertakings, if they can get hold of them, such as building, which they do now, tramways, gas, and water, upon what we may call trades unionist or even socialist principles."[155] A water company officer warned that, while other towns had been operating their water supplies economically, the LCC would not because its "administration [would be] conducted on political and not business lines. . . . The County Council have been captured by the labour party, and a good deal of their administrative work is done on what I may call political lines. . . . The London County Council have fallen, have succumbed to the temptation, other corporations have not."[156]

Preserving the Status Quo with the "Thames Storage" Scheme

Competing with the Welsh scheme for the royal commission's nod was a rival plan for increasing the volume of London's water supply: a plan to build massive reservoirs west of the metropolis. According to its designer, this scheme was inspired by the Balfour Commission's endorsement of the Thames as London's ongoing water source. "The evidence given before Lord Balfour in 1892 showed that [only a portion of the available gallons] were being taken for water supply," explained the engineer Walter Hunter. "It therefore appeared evident to me," he continued, "that if some of the superabundant water was stored when the river was running with a high flow, it could be used for water supply when the river was running at a minimum flow."[157] The project was simply an amplification of the system that had served since the days of the "Dolphin," the water intake apparatus in use on the river since the 1820s. In other words, the plan was to continue to draw from the river as always, though from a point west of built-up areas. Three large pumps would siphon water from the Thames through a large sluice covered with a perforated screen. The water would then be lifted through steel mains into two vast reservoirs with a combined capacity of three billion gallons. From there it would be pumped to the works of three companies for filtration through sand. Parliament had authorized the cooperation of the three companies and the first installment of this new system, the "Thames storage scheme" or "Staines scheme," in 1896.[158] The companies then began advocating the expansion of their system before the Llandaff Commission.

The water companies offered the reservoir plan as evidence of the Welsh scheme's superfluousness. "In the Thames Valley . . . there is no doubt that [the joint companies] would be able to overtake any demand made upon itself for an increase in its . . . supply," assured their representatives.[159] This aspect of the plan provided a counter to one of the Welsh supply's chief benefits—its abundance. The water companies' representatives also touted the fact that the Thames storage scheme could be expanded as London demanded more water; companies would simply purchase more land bordering the Thames and build more reservoirs as needed, while the Welsh scheme's watersheds were set in alpine stone and could not be expanded. The companies' agent explained that they could build their system in a piecemeal fashion, with lower expenditures and quicker returns: "Ours being a graduated scheme, it will begin to fructify from point to point very much more quickly and very much more easily."[160] The companies' Staines scheme, in the end, was much less expensive than the LCC's plan, as shown in diagrams that compared the costs side by side. The companies expected their reservoir scheme to cost £15.5 million, or £20 million to £30 million less than they reckoned the Welsh scheme would cost.[161] In placing all their reliance on the Thames, the companies also expected a great deal of predictability from the regional climate and water levels in the Thames watershed.

The water companies, too, could not rely on their continued long-term survival given that a main object of the commission was to consider their replacement. The Staines scheme provided advantageous circumstances for their buyout. The companies were literally invested in the Thames; their fates were connected with it. The Staines scheme, with its promise of abundance, was meant to make the Thames more attractive to its customers and the companies' future buyers—all potential buyers except the LCC, which repeatedly condemned the Thames. The Staines scheme was also like a savings account for the companies in that it offered them a way to build, to increase their capital in affordable, incremental efforts; they would recoup their investment and even earn some profit when they were bought out.[162] The affordable Staines scheme might also obviate the need for a large central authority to buy out the companies because it precluded the need for an expensive alternative; avoiding the buyout was the companies' ideal goal. The Welsh scheme, by comparison, was of such a physical and monetary scale that a significant governmental body would be required to finance and manage it. Alexander Binnie had testified that because London's future water suppliers "would have to go outside for a very large expenditure, I think that it had better be placed in the hands of a public authority. . . . It should not be subject to the differences of opinion among the eight companies."[163] In sum, the Thames storage scheme, by being endorsed by the Llandaff Commission, could eliminate the possibility of the worst outcome feared by the companies: the scenario in which the LCC would introduce a rival alpine water supply.

While the companies pressed forward with their argument for the Thames storage scheme, citing its affordability and expandability, they were vulnerable to the LCC's countercharge that the Welsh scheme could provide water of unmatched cleanliness, and the county council took the opportunity to press this aspect of their case. Alexander Binnie touted the Welsh water before the commission, calling it "undoubtedly a very pure and uncontaminated supply."[164] He contrasted it with the Thames supply, pointing out that in recent years there had been as many as forty-four days when the water companies had shut their intakes and been unable to draw water from the Thames because it had been so loaded with bacteria. Such evidence, he testified, "is a very serious matter for our consideration when we know that the water that came to those [intakes] flowed down from a populous district of over a million people."[165] The representatives of the companies countered that the Welsh waters were merely as pure as the Thames water after filtration.[166] "A good deal of it will contain a large proportion of peat, and peat requires more filtration than anything else," they argued.[167] On the other hand, the companies claimed that they had repeatedly proven their water's purity.[168] Professor James Dewar, an apparently nonpartisan chemist, lent support to the companies' argument. He testified that the companies' filtering made Thames water as pristine as the Welsh water, saying that it "would do equally no damage to the community, whether it is the filtered Thames water or whether it is the water from an area on which no human being is residing."[169]

Outside the Llandaff Commission's Guildhall meeting room, Hunter, the chief designer of the Thames storage scheme, appealed to the commission to reject, on economic and political grounds, the Welsh scheme in favor his plan. In a speech he gave while the Llandaff Commission was meeting, Hunter said that his plan was economical, prudent, and, since it could be undertaken by the water companies, had the virtue of upholding "private enterprise," which was "the foundation of the greatness of Britain." On the other hand, the Welsh scheme "must result in heavy additional expense," which would fall upon "the water consumers of the present and next two generations for the benefit of posterity."[170] In addition, since the Welsh scheme had to be undertaken by the LCC, it would demand the municipalization of London's water supply, and "municipalization tend[s] to discourage private enterprise," argued Hunter, "and to fasten the socialistic ideas now popular with a certain school of politicians, the triumph of which will, in my opinion, be the commencement of the decadence of England."[171]

When the Llandaff Commission finally reported in December 1899, it seemed that the water companies' and, not surprisingly, the Conservatives' case had received a sympathetic hearing. The commissioners cited economic grounds for recommending against the construction of the Welsh scheme. Pember's tactic of demonstrating the incalculability of the grand scheme appeared effective, with the commissioners reporting that "the divergence of these estimates is so great as to make the task of deciding between them difficult and hazardous." The commissioners could point to the Thames storage scheme as an appropriate alternative based on its lower expense; the Welsh scheme "is much more costly than a supply from the Thames," the commissioners reported, and since they were satisfied with the companies' filtration systems, "it is unnecessary to incur this expense now, as the supply from the Thames will be adequate in quantity and quality up to 1941."[172] Since the royal commission appeared interested in promoting a course of action that would realize the purchase of the companies promptly but not by the LCC, it was not in its interest to endorse a project that would demand that the new authority lay out an enormous sum for a new water supply; they wrote that "the Welsh scheme involves an expenditure which must be all incurred at once, and cannot be regulated and delayed according to circumstances, like that on the Thames storage works."[173] The Welsh scheme, as far as the royal commission was concerned, was dead—the "Thames storage," or Staines scheme, was upheld.

Parliament Obstructs the County Council

At around the same time that the London County Council was promoting its Welsh water project before the Llandaff Commission, it brought a bill for the scheme before Parliament. There, the Conservative majority's animosity for the LCC was evident as they opposed a project that would empower the body whose ambitions they sought to counter. When the bill was read a first time in early 1899, the county council had just one goal: get the bill read a second time and sent to committee. There, the bill could receive a more detailed hearing, with the LCC continuing to mount its public case in hopes of inspiring a mandate by electors angered by recent water failures. Also, the continued life of the rival water scheme would likely distress the companies, possibly enough to persuade them to come to terms with the LCC. Furthermore, components of the water companies' rival Staines scheme were, at the very same time, being deliberated by a House of Commons committee, and the LCC wished to have their rival project in the committee room as an alternative.

On the Commons floor, Sir Frederick Dixon-Hartland, secretary to the Primrose League, an organization dedicated to popularizing conservatism among the lower classes, took up the standard against the LCC on behalf of the Conservatives. To his eyes, the Welsh scheme was intended to give the LCC the power to sidestep, and ultimately ruin, the water companies. "By allowing the London County Council to obtain the Welsh supply they will then have the power, if they please, to make an alternative system of water supply to that provided by the water companies," he stated in debate.[174] C. A. Whitmore, a Tory MP, seconded Dixon-Hartland's objection that the Welsh scheme would give too much influence to the LCC, saying, "I cannot conceive myself how . . . a Select Committee of this House could give its sanction to the acquisition of a great tract of country and the construction of reservoirs without being certain that body which undertakes this great task was certain to be the water authority for London."[175] Both Dixon-Hartland and Whitmore were arguing that the body that undertook the task of finding London a water supply was certain to be the city's water authority; hence, in their view, that body must not be the LCC.

Dixon-Hartland, chair of the Thames Conservancy, also had grounds on which to object to the scheme: it impugned the quality of the river's water and his board's work. "The Thames is a river which has a large population upon its banks," argued James Stuart, an MP and London county councilor, during the Welsh scheme debate. "In respect of pollution," he stated, "you will see that there is going on, in spite of the praiseworthy efforts of the Thames Conservancy, a large amount of pollution."[176] "No," interrupted Dixon-Hartland; the Welsh scheme was unnecessary in part because the Thames was not polluted, or at least not *very* polluted. "A report which was presented only last week," insisted Dixon-Hartland, "shows that only 113 cases of pollution at the present moment exist, and that water has been purified to such an extent that 500,000 people who in 1884 had polluted water now have pure water."[177] The Welsh caucus (including David Lloyd George, formerly a Welsh nationalist) offered little to the debate.[178] Their few comments revolved around a practical concern that the Welsh farmers displaced by the new reservoirs would receive adequate compensation from the LCC. "Indeed, we shall be, for many reasons, glad to welcome the London County Council in Wales. At the same time," stated A. C. Humphreys-Owen, whose constituency would receive a number of the LCC's reservoirs and works, "I hope I may be able to receive from [James Stuart, LCC member and MP] some statement as to the intention of the Council to provide as fully as they possibly can, for the small Welsh farmers whom their operations will displace."[179] James Stuart, of course, assured the Welsh representatives that the LCC would compensate all the Welsh claimants and replace any schools, businesses, and other buildings submerged.[180] The assurances appeared to satisfy the Welsh members, all of whom voted for the second reading of the bill.

The debate had gone on for about two hours when the final word was offered from Henry Chaplin, chair of the Local Government Board. The opponents of the bill took the strategy of arguing that it would be rash to give the LCC the power it wanted when the Llandaff Commission was already dealing with all of the issues of London water supply. James Stuart, the county councilor who bore the brunt of the work of defending the bill, countered that the royal commission "has certainly taken evidence on the matter a considerable time ago, but that evidence is finished; and although the Commission has not reported, that evidence can go before the Committee on [this] Bill. . . . The effect of the Royal Commission is such that it neither influences whether this Bill should be read a second time or not."[181] The Conservatives, it seemed, would take the calculated risk that the public was not so exhausted with the companies that they would be outraged by the obstruction of the LCC. "This bill . . . raises the question who is to be the new authority. . . . If the London County Council acquire the property they are seeking to acquire," Chaplin said, "that practically settles the question of who is to be the new authority" over London's water supply.[182] The House's majority knew that it had the numbers to defeat the bill.[183] After Chaplin's last word, the House divided, 206 to 130. The Welsh scheme was defeated.

The London County Council was holding its weekly meeting while the House of Commons debated its bill. "When news arrived that the County Council's Welsh water scheme had been rejected," the *Daily Chronicle* reported the next day, "there was much indignation among the majority of the members."[184] George Shaw-Lefevre said, "It is impossible to disguise the fact that the water companies were congratulating themselves very much on what had happened in the House. In point of fact the water companies got all they wanted. They prevented the second reading of the Council's Bill, which would not now go to the same committee as the Companies' [Staines scheme] Bills."[185] Other councilors tried to salvage some political gain from the defeat. The Conservatives, they argued, had shown disdain for the will of the London voters. "The unanimous decision of the Council representing the people of London to obtain a new source of supply was thrown aside," claimed E. A. Cornwall, adding, "It was a good object lesson to London to see how the interests of the people were dealt with by Parliament."[186] By rejecting a new, pure source of water, the Conservatives would be culpable if a typhoid outbreak or other epidemic occurred in London, as it recently had in a nearby town, argued Cornwall and Dr. W. J. Collins.[187]

Some councilors remained optimistic, however.[188] Alexander Binnie was hopeful that the LCC's water committee could regroup and resubmit the Welsh scheme proposal. The failure of the bill, he suggested, "afforded us the means of testing our weak points, feeling where opposition would be most severe and generally giving us the opportunity of improving the whole project."[189] And, indeed, in 1900, the committee did consider resubmitting a bill, asking the engineer and parliamentary committee how much their proposed plan would cost.[190]

In April 1899, W. C. Steadman, a Fabian MP for East London, summed up the blow to the LCC represented in the Welsh scheme's loss, saying in the House that, "from the action of the present Government in throwing out the London County Council's Wales scheme, I know that we cannot expect any proposal from them for the purchase of the water companies' undertakings at a fair and reasonable price, vesting control of the supply in the London County Council."[191] He was right. The Welsh scheme was in fact finished and never to return, and lost with it was a key strategy for securing the water supply.

The new water supply projects that proceeded in the wake of water municipalization in dozens of British towns from the 1840s to the 1890s were works of environmental change intended to realize social change. The new water sources were responses to the practical problems of the industrializing city, to be sure, but in the eyes of their designers and promoters, they were means of realizing a new kind of moral local government, an elevated civilization, and a new way of life, especially for the poorest members of society. The London County Council's Welsh scheme was to be a work of environmental change for the purpose of social change, but on a scale not pondered by earlier reformers. The LCC shared the goals of those forebears, but its Fabian-drafted platform called for an even deeper modern-moral transformation. In the Progressives' eyes, the local government needed to take more direct action in citizens' lives, change the basic arithmetic of labor, wages, and taxes, and uproot the power holders from their entrenched place in the status quo. The LCC, as guided or goaded by the Fabians, sought structural changes for which no local government had ever strived, and changing the way water—the first necessity of life—was obtained was an integral part of this objective.

The LCC's grand ambitions were hardly welcomed in every corner. Conservatives argued that the council would at the very least hurt middle-class Londoners with increased taxes. At the worst, the Progressive-dominated council might lead a bloodless revolution against private property, which Conservatives held as inviolable. They saw the LCC seeking to gather more and more powers in its own hands so that it might some day come to rival the national government in wealth and influence. Water municipalization had carried on at a steady pace throughout Britain for half a century, but by the LCC's second term, Conservatives counseled keeping water out of the hands of this new local government. Before the Llandaff Commission, the Conservative MP C. A. Whitmore was asked why he opposed placing the water supply in the hands of the LCC when throughout the century a number of commissions and committees had recommended bestowing the responsibility for water service on a local authority. Whitmore responded that it was because "all of those reports . . . were made before the County Council developed characteristics which are . . . peculiar to itself."[192] And so the national government blocked London's project of environmental change, even though London's long-standing water sources were failing at the time under extreme weather conditions. Authorizing the Welsh water plan would have made the LCC the master of London's water supply and would have granted them a keystone in their wider strategy for London.

A recurring motif in the literature on modernization and the land holds that the burgeoning central state sought to "modernize"—as its members understood the term—by undertaking massive projects of environmental change or rationalization.[193] This case offers an adjustment to that picture, or rather, it presents another process of modernity. In Britain in the nineteenth century, the greatest government projects aimed at environmental-social change were undertaken by disparate local governments. The case of the Welsh scheme introduces a further complication to the traditional picture. First, the central government blocked this project. The scheme represented a way to realize an improved society just as the Loch Katrine, Thirlmere, or Vyrnwy works had, but, in the way the Welsh plan served the LCC's vision of a new kind of city, it also represented a tendency toward collectivization. Second, the central government supported a project proposed by corporate interests that was the reverse of the type of water scheme that had been built for decades. The projects undertaken by the provincial authorities were intended to introduce water of unimpeachable purity and to secure, in a stroke, water sources of such volume that future generations would thank them for their foresight. They were designed, in other words, to represent the high purpose of a new generation of moral governors.

On the other hand, the Staines scheme endorsed by the Llandaff Commission and later authorized by Parliament drew its water from the same source that London had always called upon: its reservoirs wedged between villages just upriver from the metropolis. It was a plan conceived not by the local representatives envisioning a modern-moral civilization but by profit-seeking companies. These private firms were not securing a new water source to safeguard many future generations at a stroke but were instead trying to increase the volume of their product with the least possible investment and expense to themselves. The rival version of modernity it was meant to promote rejected the expense, activism, and centralization of the Progressive London County Council.

The weather had intervened in the contest over London's water supply, with residents suffering through droughts unseen for several decades. The reputations of the water companies, already viewed negatively among Londoners, reached a low point. The companies' unpopularity was so great that Salisbury's government was risking electoral revenge by obstructing the LCC's efforts at water reform. It seemed that it had to either allow the LCC to proceed or—for the first time—directly intervene and take in hand London's water reform.

⇒ CHAPTER 5 ⇐

An Alternative Vision of the Modern City, an Alternative Government of Water

ROM THE MID-1890s into the new century, the Salisbury government clearly understood that to control the flow of water was to control the flow of power. In their dealings with the London County Council's attempts to take over water service in the capital, the Conservatives who controlled Britain's government proved themselves deft practitioners of the craft of water politics. They obstructed the LCC's bills seeking power to purchase London's water companies, they opposed the council's Welsh scheme that would have served the LCC's vision, and they supported the rival Staines scheme that facilitated private water control. For the last decade of the nineteenth century, the Conservative government's strategy in dealing with the LCC was simply to thwart its attempts to remake the city through the control and manipulation of water. As the nineteenth century drew to a close, however, the government found that a strategy of negative reaction would no longer suffice.

Drought overtook the region, and the water companies that the Conservatives had been protecting simply could not meet London's needs. London's Conservative MPs and the Moderates on the county council feared electoral reprisal. The voters demanded action, but the Conservatives could not let the LCC triumph. In their eyes, the LCC was led by doctrinaire, ambitious, acquisitive politicos who, if given the chance, would abuse democracy and seize control of land, water, and industry. In these years, Conservatives and dogmatic Liberals were beginning to envisage a future in which private enterprise and property were threatened by the spread of municipal activity. They could not allow London to become a vanguard of collectivization and centralization. They believed the government needed to take action and install its own water administration for London—the Metropolitan Water Board (MWB).

Conservatives recognized that a system of environmental administration could influence systems of power. It could reinforce social structures, support economic trends, and obstruct others and could also create authority where none had been before. The leaders of the national government carefully crafted a water administration designed to distribute power where they wanted it in order to realize their own vision for the future of London. In this London, state enterprise would be kept to a minimum, the centralizing tendencies of the LCC rejected in favor of more localized control, and volatile democracy tempered with altruistic oligarchy. Gas, water, locomotion, labor, housing—the core elements of the urban ecosystem—should not fall into the hands of the Fabian and Progressive puppet masters who would manipulate these services for their own revolutionary ends. Instead, they intended control to be placed in a mix of private hands, corporate hands, and, frankly, Conservative hands. This was a modern London to oppose that of the Fabians.

Linking Municipalization with the Spread of Collectivization in 1900

In 1899, a group of influential Conservative politicians began to campaign for an official parliamentary inquiry into the growth of municipal enterprise throughout Britain. They pointed out that Parliament would consider in the upcoming session—along with the LCC's water bills—approximately seventy bills for the establishment or municipalization of various utilities and services, from water to electricity to the manufacture of boilers for civic operations.[1] This surfeit was, in their eyes, cause for alarm. "Municipal trading," wrote the Earl of Wemyss, a longtime MP and formerly a member of the Liberty and Property Defence League, might come to "supersede all private trading." He considered this prospect a threat to "human progress" itself. He called for a joint committee of the Houses of Parliament to "put a stopper on municipal ambition and the speculating of municipal authorities with the ratepayers' money."[2] John Lubbock, now Lord Avebury and a former Moderate member of the LCC, joined Wemyss in campaigning for parliamentary hearings, which the Commons and Lords agreed to the next spring, even while the LCC was introducing yet another purchase bill.[3] And so in May 1900, the Joint Select Committee on Municipal Trading began to take evidence to determine whether municipal enterprise had expanded too far.[4] Witnesses included industrialists, small manufactory owners, municipal figures, town clerks, and other experts.

Queries tended to aim at discovering the hidden threats and dangerous trends encouraged by municipal enterprise. Could municipalities run their operations in an efficient manner? Did municipal enterprise increase taxes and city debt? Did they unjustifiably undermine commercial enterprises?[5] The discussion often returned to the question of corruption. Would town councils turn municipal operations into the cogs of a political machine, directing large voting blocs of employees in return for financial rewards? On the other hand, would town councils bend the law in order to favor their enterprises over commercial rivals?[6] Municipal governments could be tempted to make a profit at taxpayers' expense, suggested the committee's chair, while another member hinted that municipalities would be too willing to take a loss on their operations, since they were merely spending other people's money.[7] Other lines of questioning raised the likelihood that municipal enterprises tended to increase their payrolls, to drive up wages for private commercial interest, and were imprecise with their financial bookkeeping—even to the point of fraud.[8]

The joint committee called in representatives from the LCC, including the chairs of its housing and transportation committees, to inquire after the financial soundness of the LCC's activities and whether or not private enterprise deserved some security against public ventures and to examine their opinions on how far municipal enterprise should be extended.[9] They heard the testimony of Lord Avebury, too, who stated, "I have been alarmed to see just how far the advocates of the present system would go." For him, the principle followed by municipal leaders for more than half a century—that it was the duty of local government to take under their control the mechanism for maintaining the urban environment—had been revealed as mistaken. "As regards water and gas-lighting, and tramways . . . fresh undertakings should only be approved if it can be shown that there are special reasons in the particular case why they should be undertaken by the local authority rather than by private enterprise," he stated.[10] Parliament, he concluded, should intercede by inspecting more closely each bill for transferring formerly private enterprise to municipal control, which Parliament had tended to approve without difficulty in the past.[11] The central government, put another way, should intervene in the interest of liberty.

Nothing grew out of the Joint Select Committee on Municipal Trading in terms of practical legislation. The close of the parliamentary session drew it up short and it simply published its proceedings, but the episode revealed the sense of apprehension among leading Conservatives at what the modern city might become. They saw a possible future in which the activity of grasping civic governments would depress private enterprise, lead to an indolent nation, and leave only corrupt and bloated municipal kingdoms. Conservatives had sat back for too long as innocuous gas and water socialism had turned into something else.

The Mounting Risk of Inaction

In mid-September 1898, as drought gripped London and taps went dry once again, protesters filled Trafalgar Square, all clamoring for the elimination of London's water companies. A week later a group of Unionist London MPs went as a delegation to the Salisbury government's Local Government Board to appeal for the public takeover of the companies, despite their political aversion to interference with commercial enterprise. Meanwhile, public meetings of Conservative, Liberal, radical, and socialist organizations alike turned their ire on the water companies while expressing their sense that the government needed to intervene.[12] In the fall of 1898, one of London's parish vestries circulated a survey among all of the metropolis's thirty-eight vestries and district boards inquiring whether the bodies favored placing the water supply in the hands of a public authority and, if so, what form that body should take. Of the thirty-two local bodies that responded, twenty-eight favored placing the water in the hands of a public body, and a slight majority of those made clear that they wanted the LCC itself to take over the water supply. There was some tendency for wealthier western and west-central vestries—those that had escaped water stoppages—to favor the status quo or not to reply to the survey, but, for the most part, opposition to the companies, a preference for public control, and an explicit demand for the LCC to administer the water supply were spread evenly across the metropolis.[13]

In this context of public anger directed toward the water companies and electoral calls for the government to take action, the London County Council promoted its Welsh scheme and purchase bills in the spring of 1899. Though these bills failed, the environmental factors kept the pressure on the companies and the national government as drought returned once again in the summer of 1899. "Providence," argued the devout W. H. Dickinson in LCC debate, had "intervened to show that the Council's proposal for a supply from Wales is the real remedy."[14] In reaction, the LCC voted its water committee a £3,000 budget to prepare water bills for the upcoming session of Parliament. The council subsequently took its purchase bill back to Parliament in 1900, only to have the Conservative majority block it once again.[15]

As water shortages recurred, a policy of mere obstruction invited electoral reprisal against both the Moderate Party in the LCC and the London MPs. The Conservative parliamentary majority, after all, had blocked LCC purchase bills almost yearly ever since coming into office in 1896. That year, the Salisbury government created its own cabinet committee to wrestle with the water question. It consisted of Lord James of Hereford, a former attorney general, as well as Henry Chaplin, head of the Local Government Board, and Joseph Chamberlain, who had guided gas and water municipalization as mayor of Birmingham before entering Parliament and leaving the Liberal Party. The cabinet committee's memoranda show that the Conservative government's Moderate allies on the LCC and the Conservative London MPs grew concerned in the second half of the decade about appearing obstructionist. In 1897, Lord James wrote to the other members of the Salisbury government's water triumvirate that

"the Moderates" not unnaturally take into consideration not only what is best for London, but also what effect any policy will have upon their position with the municipal constituencies. They also have a general desire to oppose the action of "the Progressives." Their belief is that London is strongly in favour of "something being done" to take control of the Metropolitan Water Supply out of the hands of the Water Companies. Therefore they say if "nothing is done," the Moderate Party and the Unionist Parliamentary Party will be seriously injured in the judgment of the London Municipal and Parliamentary constituencies.[16]

Later that year, the Moderates and the cabinet committee agreed to institute the Llandaff Commission as a way to take a degree of visible action without committing to any course. In the meantime, the county council might change hands, and, freed from the immediate pressure of public agitation, the government could formulate its own solution to the administration of London's water supply.[17] In short, a royal commission provided grounds for the government's inaction until it reported.[18]

Decentralizing the Center

In the meantime, the Salisbury government was making a more direct assault on the LCC, with great consequences for the future administration of London's water supply. Just before returning to the post of prime minister, Lord Salisbury had called the London County Council "the place where a new revolutionary spirit finds its instruments and collects its arms," and in a speech in fall 1897, he intensified his denunciation. Speaking before an audience of Conservative organizations in Royal Albert Hall, the prime minister first railed against the LCC's Progressives, saying, "You should get rid of people who have such an exaggerated ideal of civic duty. . . . The statesmen of this county have fallen victim to a common intellectual complaint of the present day which . . . I may name megalomania—the passion for big things simply because they are big." Then Salisbury turned his invective on the body itself, stating that "London has departed from the tradition and precedent of all other large municipalities, and has got a constitution wholly different from theirs. . . . You have got a little Parliament—and a Parliament is not what you want. . . . I think that is a proceeding destitute of either wisdom or judgment." These hardly resembled the words of a party leader hoping to motivate Moderate candidates to capture the council—and indeed the Moderates were routed in the 1898 council election—but Salisbury had a different tactic in mind for reining in the ambitious LCC. In his speech, he declared his ministry's intention "to give a large portion of the duties which are now performed by the County Council to other smaller municipalities."[19] The Conservatives intended to reduce the council's powers and so shift responsibilities from the LCC to new second-tier authorities akin, as Salisbury stated, to municipalities themselves.

A few months later, a committee of Conservative vestry members and LCC Moderates met to formulate these new bodies and consider the manner for parceling out some of the LCC's power among them. The committee advised that Parliament create twenty-four "municipal corporations," which would replace the vestries and assume their duties. The act creating these corporations should provide a procedure through which the bodies could apply to the LCC for additional powers to be transferred down to them; if the LCC refused, the corporations should be able to appeal to the Local Government Board for the powers. The committee trusted that a natural desire for local autonomy would result in widespread demand for significant transfer of power.[20] The Local Government Board, too, should have the power to grant the new bodies loans so that the second-tier authorities would not have to rely on the LCC for funds as had the vestries. Most importantly, in terms of enervating the county council, the mayors of the new corporations would serve as aldermen on the LCC. To a degree, this would give the council something of the character of the defunct Metropolitan Board of Works—a congress of delegates sent from widespread local bodies with widespread local interests.[21]

The LCC's Progressives, naturally, reacted strongly. "If [only] the desire for reform," wrote Dr. W. J. Collins in the *Contemporary Review,* "rather than jealousy of the County Council had been the inspiration of the Bill." He warned that "the disintegration of the growing unity of London into a conglomerate of sham municipalities . . . will postpone indefinitely that unity, simplicity, and equality of treatment which are the cardinal principles of the reformation of London."[22] The Progressives, of course, condemned the devolution scheme on the floor of the county council, while newspapers sympathetic to the Progressives also took up the cry.[23] The radical *Star,* for example branded the campaign "Salisbury's war on the LCC" and the Liberal *Daily Chronicle* pointed to government leaders' "cordial hatred" of the council as the motive for the devolution effort.[24]

The campaign also had supporters in the public discourse. The editors of *Blackwood's Edinburgh Magazine,* perhaps exaggeratedly, described the Conservative plan to shift power away from the LCC as a possible preventative against communist revolution. The Progressives, the magazine pointed out, sought to employ not only road and other works crews but also utility providers and even the police. "The County

Council, pledged to a policy of communism, would have a great army of . . . the labouring population," the magazine editorialized. "Such are the beginnings of all revolutions," it continued, "and it is this consideration which forms the strongest ground of all for the new London Government Bill."[25] The pro-Conservative *City Press* was more measured, supporting "the municipal movement" as a counter to the Progressives' efforts "to magnify [the council's] powers beyond reason, and use it for the furtherance of all kinds of socialistic purposes."[26] The *Times* was a bit more measured, arguing that boroughs in London would be operated as "provincial towns have been, on the whole, efficiently, economically, and honestly," with the LCC's effectiveness providing the unspoken counterexample.[27] And the water companies, naturally, made clear their support for devolution because it had the potential to eliminate the LCC as a viable purchaser of their businesses.[28]

The bill that finally came down from the Salisbury government in spring 1899 was hardly the revolutionary decentralization of power for which the LCC's critics had hoped. Most vestries were uninterested in assuming greater responsibility, and the 1898 LCC election, which the Progressives won in part on a platform of deriding Salisbury's devolution offensive, demonstrated the unpopularity of the anti-LCC campaign.[29] "[The] Bill is not on the lines of the scheme which the London Municipal Society [the Moderates' electioneering organization] proposed. It wanted separate municipalities and diminished authority for the County Council, neither of which it got," reported the *Municipal Journal*.[30] In its final form, the bill did not call for the creation of new municipalities but for twenty-eight new second-tier units called "municipal boroughs." These entities would not assume the lion's share of duties in the metropolis—in fact, they were to assume only a handful of duties beyond those of the old vestries—nor did the bill include a mechanism that would easily allow them to draw powers down from the council. The London Municipal Society sought to re-intensify the bill with amendments, but to no avail. Despite continued opposition from the LCC and its allies, the government secured its London Government Act of 1899, even though it was incapable of rapidly crippling the county council.

Though Conservatives expressed disappointment with the measure, the creation of the municipal boroughs would soon have immense consequences for the struggle over London's water supply. Opponents of the LCC saw a way to use the new authorities to derail the Progressives' ambitions. The origins of the idea are uncertain, but in 1896 a Marylebone vestryman wrote that since "the metropolitan water question should be dealt with next Session, may I . . . suggest the eminent desirability of dealing with the subject of London Government at the same time? These two matters have a very close relationship. . . . The control should be in the hands of . . . new local corporations, and not in those of the London County Council. . . . Many . . . view with the keenest distrust any further increase in the County Council's powers."[31] The next year, one of the leaders of the devolution movement also promoted the idea. Lord Onslow, a leading Moderate on the LCC who was also undersecretary of state for India, wrote to Salisbury of the potential for more powerful second-tier authorities to assume control of the water supply in lieu of the London County Council.[32] Even if popular opinion favored public control of the water supply, the power of water could be decentralized, at least—kept from the hands of the acquisitive Progressives.

The Emergence of a Government Plan

The Royal Commission on Water Supply finally reported in December 1899 after having met for two and a half years—"prosecuting its inquiry in a leisurely manner," in the words of one county councilor.[33] It recommended, like many authorities before it, the elimination of the private water companies in favor of public control. A public water authority, the commissioners concluded, could address London's future water needs—as well as the difficulties created by the recent series of droughts—with coordinated action and with an economy that would benefit taxpayers.[34] But the commission rejected the LCC itself as the appropriate water authority; it also rejected the LCC's scheme for securing a future supply in favor of a more conservative engineering plan. Instead, it recommended a thirty-member water board to undertake the purchase and administration of London's water; the LCC was to provide ten representatives, the Thames Conservancy Board four, the River Lea Conservancy Board two, and two each from the five counties and one municipal corporation also served by London's water companies. The government would appoint the board's chair and vice-chair, who were to provide expert knowledge of the water trade and engineering and lend the body a degree of continuity.[35]

The commission's recommendation followed closely along the lines of a policy the Salisbury government had briefly pursued in 1896. The LCC strongly opposed the idea then, with W. H. Dickinson, the one-time LCC water committee chair, for example, telling a public meeting that the government bill "merely creates a novel and unnecessary body . . . as an effective pretext for killing the County Council's bills." Sidney Webb rallied the same audience to spread opposition to the Conservative policy through "small local meetings up and down London; by taking every opportunity of addressing even a dozen or twenty electors in the remotest suburbs."[36] And the surrounding counties showed little enthusiasm for the bill, being concerned more with protecting their own water sources than with the management of the metropolitan water supply.[37] The 1896 bill languished in the Commons. With the inauguration of the municipal boroughs in 1900, though, the political landscape changed dramatically from what it had been in 1896, and the central government saw a way to resuscitate the idea of a board in order to keep water out of the hands of the LCC and to strengthen the new local bodies in one stroke. "The constitution of the Water Board proposed by the Royal Commission does not appear to me to be satisfactory," wrote the president of the Local Government Board to the Salisbury cabinet; the commissioners had not included the new borough councils in its proposal. Walter Long recommended giving each borough a seat on the board, while granting the LCC only five seats.[38] The government concurred. It settled on a firm course of action.

The central government had at long last intervened decisively in the long debate over London's water service. The control of water must be centralized, the national government determined, in order to save society from pernicious centralization, that is, the Salisbury government determined once and for all to place water service into the hands of a trust in order to keep it from the hands of an authority with collectivist aims. The move was to ensure that water centralization must not mean water modernization as envisioned by London's Progressive leaders. The national government was ready to bring its bill forward in early 1902—when the water shortages were still producing calls for action, when the Llandaff Commission had already reported and shown that private companies could not satisfy London's water needs, and when the Conservatives were enjoying a large majority in the Commons after their great success in the Boer War–influenced election of 1900 in which they won fifty-two of sixty London constituencies.[39] The LCC's

FIGURE 5. The black line on this map of the County of London (shaded area) defines the area served by the metropolitan water companies. Adapted from Richard Sissley, *The London Water Supply* (London: Scientific Press, 1899), 13.

Progressives, too, provided pressure for action, having increased their majority in a 1901 election run largely on the question of water municipalization—a fact hardly lost on the council's opponents in Parliament.[40]

The government's 1902 bill called for a water board of sixty-nine members, with the LCC entitled to only ten of them. The balance of the membership was to be drawn from a wide area and a range of authorities. The largest bloc was to come from the metropolitan boroughs, the majority of these sending one member each but the five with the largest populations sending a pair. The urban district councils (the local sanitary authorities for London's far-flung suburban areas) were to send eleven members—the next largest bloc. Four neighboring counties were to send one member each, the City of London and the City (metropolitan borough) of Westminster, two each, and the conservancy boards of the rivers Lea and Thames, one member each.[41] The board would elect a chair and vice-chair, envisioned as water experts, from the outside. The Metropolitan Water Board was to have the power to levy rates just as the companies had done, calculated as a percentage of the rental value of each premises. In the case of revenue shortfalls, the water board could levy the local authorities represented on it. The bill called for the companies' undertakings to pass into the board's hands within an appointed period, with purchase negotiations taking place after this point, if necessary. It named a court of arbitration to mediate the purchase: a retired lord justice of appeal, a retired Local Government Board secretary, and a prominent engineer.[42] The companies would be compensated with water board stock secured on the companies' property and with interest serviced by future water rates. Following the model of a political campaign, the president of the Local Government Board organized a group of Conservative London MPs to encourage support for the new board among London electors. These electors, presumably, were to pressure their borough councils to demonstrate an interest in taking part in London's water administration.[43]

The Central State's Rival Government of Water

The new system of municipal borough councils administering water supply served the national government's interests principally for the simple reason that it kept the water supply out of the LCC's hands. Walter Long, the president of the Local Government Board and leader of the Salisbury government's water campaign, in subsequent years explained the appeal of the water board bill by writing that "the advanced views of many members of the Council gave rise to feelings of alarm at entrusting them with further powers."[44] Water, in other words, would be saved from the same fate as the cities' tramways, for example, which fell into the Progressive-dominated LCC's hands in the same years, and as the terms of labor, handled through the LCC's large works department. The new bill could eliminate, too, the potential for the income from water rates to fund further LCC endeavors, as had been the case in towns that had municipalized companies previously.[45] The Metropolitan Water Board's constitution promoted the weakening of the county council by empowering the borough councils created to administer water service. If they could be granted more collective authority over London's water supply than the LCC enjoyed, the potential existed to grant them further powers that would supersede the LCC's authority in the future. The LCC's Progressives, even if they could send representatives of their own party to the board, could control at most 7 percent of the Metropolitan Water Board's vote. The representatives of the municipal boroughs were hardly likely to ally themselves with the council; as the Conservatives had anticipated, the party held control in the majority of the new bodies after the first borough council elections in 1900.[46]

The proposed Metropolitan Water Board represented a break in the history of urban modernization through water supply administration. Most simply, municipalization by trusts or boards had been rare since the beginning of the movement in the 1840s and 1850s. At the height of the trend in the early 1880s, more than 90 percent of municipalized water supplies were owned and operated solely by the chief city or town authority.[47] The others were administered mainly by joint committees comprising representatives from a handful of town or borough councils that united to purchase commercial undertakings and build waterworks jointly. In the northwest of England, Ashton-under-Lyne, Stalybridge, Dukinfield, Mossley, and Hurst—representing a combined population of approximately one hundred thousand—joined in 1870 to build a gravitation scheme devised by John Frederic Bateman.[48] Edinburgh, Leith, and Portobello, too, fell into this category.[49] In other cases, towns lacked a suitable local authority to administer a municipalized water supply so they instead invented a water commission.[50] Shared water supplies were not always administered by joint boards. There were important precedents for cities municipalizing companies that had served wide areas and then selling water back to outside areas without giving them representation. Manchester, for example, purchased companies that served neighboring areas, and when it operated its own supply, the city simply sold water in bulk back to Salford and many other towns.[51]

Opponents of the government's plan offered many more examples of cities municipalizing water supplies and then selling water to outside authorities without providing them representation on their commit-tees.[52] And the national government never demanded that local bodies sharing a water supply unite in a water board as a condition of granting an act for municipalizing water companies. The national government, however, justified the requirement that outside bodies be represented on a London water board by arguing that those areas would be de-pendent on the LCC for water but the areas' interests would not be represented in that body. "While the London County Council rep-resents inner London, it does not represent . . . that large outer London which is so vitally concerned in the proper and wise settlement of this question," argued Walter Long.[53] "As the population of 'Water London' grows," he added, "it will be in the outside areas that the great increase will take place."[54]

A particular set of ideals drove the leaders of towns and cities that municipalized their water supplies in the second half of the nineteenth century. "It is the duty of a wise local government," argued individuals like the Birmingham borough councilor Thomas Avery, "to endeavour to surround the humbler classes of the population with its benevolent and protecting care." For Avery, doing so meant "the abundance, per-manence, and regularity of the supply of good water."[55] Government had the duty to protect citizens from one another in terms of public health, had the duty to make possible citizens' human dignity and mo-rality in the face of degrading living conditions, and had the respon-sibility to protect citizens from the lurking danger of privation. These ideals had grown stronger in reaction to an environmental crisis that threatened all of them, a crisis that gripped rapidly growing popula-tion centers starting in the first half of the nineteenth century. In 1902, the national government intervened in London in reaction to environ-mental pressures as well, amalgamating the water companies' supplies and distribution systems partly for the purpose of preventing water shortages, as the Local Government Board president Walter Long ex-plained.[56] The water companies' position had become untenable.

But the similarities in motivation for the towns and boroughs ver-sus the national government do not extend much beyond practical con-siderations. Provincial leaders who had succeeded in municipalizing their water supplies had spoken of building a new civilization of health and abundance for many generations to come, of a new kind of local government that made the moral life of its citizens its concern, and held it as their duty to reflect the superiority of British civilization in the outward appearance of British cities. The figures within the Salisbury government did not express such lofty rationales for transforming the government of water. They did not hold the kinds of stakes in London as provincial town councilors held in their cities—they did not see themselves in the role of London's improvers.

For provincial water reformers, the principles on which the administration of water was based—as well as the engineering principles on which water provision was based—were meant to make their cities more modern in the sense expressed by Avery, the Birmingham councilor. Water administration by a directly representative body was to provide an obvious contrast to the commercial companies that made independent decisions about water quality, abundance, and price based on the profit motive. "When water is under the control of private companies, the chief desire of the directors is to obtain good dividends," said a Bradford town councilor in 1852. "When the Town Council possesses the works," he continued, "their chief object is to make the works instrumental to the promotion of cleanliness, the health, and the comfort of all classes of citizens."[57] And the municipalizers' technology—their new vast waterworks—were meant to bring about fundamental social change, as well. Gravitation schemes reflected a demand among urban leaders for the highest quality water possible for all classes of citizens, with water drawn from what were supposed to be the hinterland wilderness landscapes that were a tonic for city dwellers. With their integral high water pressure, the systems were to deliver water in abundance to every nook and cranny of every dark court and alley of the city. Scattered public pumps were to be banished; landlords could no longer plead inability to offer water to the upper floors of their tenements. The poorest townspeople were to learn new habits of cleanliness.

When the initiative of the water reformation moved from multiple urban points to the single central state, there were no such expressions of idealism in the plans. The intention was in large part to keep the influence of a partisan electorate out of decision making about water collection and control. Under its scheme for a board, the London constituency—a constituency the Conservatives in the national government had learned to distrust since 1889—could not easily determine its water affairs. If Londoners disapproved of a water supply project, for example, if the board mismanaged the supply so that London experienced shortages, electors could not replace the board's personnel in a single election as they could with the LCC. At most, Londoners could replace the water board's ten LCC representatives at a single election, waiting for metropolitan borough council elections to replace thirty-two more, with the outlying urban district councils, county councils, conservancy boards, and so on, determining the remainder of the sixty-nine seats. The board's leader, a paid expert, was to have no constituency at all. The Metropolitan Water Board, in sum, was responsible to no single electorate. The government's plan for the future provision of London's water had none of the integral modernizing features of the many massive projects undertaken by provincial capitals. The "Thames scheme" endorsed by the Llandaff Commission ensured that the river would go on supplying London as it had for centuries. It did not provide for a level of water quality above that provided by the water companies, nor did it, as a scheme that continued to rely on steam pumping, provide high-pressure water to all corners of the metropolis at all times.

Protests against the Central Government's Move

The water companies, naturally, did not suddenly adopt an interest in their own demise simply because the Conservative government now declared that their demise was ensured. Edmund Boulnois, chair of the West Middlesex Water Company and Tory MP for Marylebone, opposed the government's intentions in the House of Commons. The idea of compulsory purchase, he argued, was "unjust and unreasonable." "It cannot be said with any justice that the companies are in default [due to water failures]," he said, adding, "On the contrary I maintain that the companies deserve the thanks of the community. They brought water to London many years ago, when no one else either would or could."[58] And Boulnois did receive a sympathetic hearing in some quarters. Colonel A. R. Mark Lockwood, a Tory MP, argued that "the principle of purchase, although, I am opposed to it . . . has been growing among Members of the House for some time, and it is now practically an accepted theory." On the other hand, Lockwood thought that the Conservative government was making the move only reluctantly, and, lacking any desire to "confiscate" property and interested in "dealing fairly," it was in his opinion the appropriate body to undertake the action. He urged the government to be as generous as possible to the companies' shareholders out of respect for private property and to discourage any precedent for the easy public acquisition of private industries.[59] But with the Conservative government resolved to bring to a close the London water question in 1902, the companies lost their parliamentary allies of ten years.

Major Frederic Carne Rasch, a Conservative MP who chaired a water company outside of the metropolis, counseled his fellow water company directors to accept the government's proposal, arguing that the stockholders "ought to be glad to throw themselves into the arms rather of the Local Government Board than of the London County Council. . . . If they wait long enough it is absolutely certain that the London County Council will come in and decide the question in a way the water shareholders will not like."[60] Stockholders truly had two reasons to be wary of the LCC. First was the possibility that the council could win approval of its Welsh scheme—perhaps after the 1906 election—and offer a competing supply. Second, the LCC had argued for years that it would not pay the companies the straight value of their stock but would take into account the fact that their operations were likely to prove insufficient in the near future, requiring a large investment in new supplies.

The government's intentions for water were naturally unacceptable to the London County Council's majority, despite the fact that London's water was finally to be divested from the eight commercial suppliers. For the Progressive council members, water should have a positive role in the new kind of city they envisioned. That vision owed much to the Fabians, and Sidney Webb had rejected the idea of placing London's water in the hands of a trust when he had offered a picture of London as an ideal city in his *London Programme*, released in 1891.[61] He sought the removal of water from the hands of private water rentiers, just as he wanted land removed from the hands of private capitalists. In a positive sense, he wanted the income from water sales to strengthen the local government so that it could, in turn, enact further municipalization. He wanted the construction of waterworks and the daily operation of the water supply undertaken by civic employees so as to increase the scope of municipal labor. In terms of principle, he believed that the ideal city should make communal decisions about the provision of life's prime necessity. Under a trust or board, the decisions would be made out of sight, where the electorate had no access. And so, from the time the LCC digested the Llandaff Commission's report, the council's Progressives spoke out against the idea of the Metropolitan Water Board on the grounds that it enfeebled the county council and violated what they called "modern" principles of good local government.

In response, an LCC deputation led by W. H. Dickinson, chair of the LCC and former chair of its water committee, called on Walter Long, president of the Local Government Board, to protest the national government's unusual step of demanding the formation of a board representing all areas that were to receive the amalgamated water supply. If the Salisbury government believed it was stepping in to safeguard hinterland areas' independence, the Progressives were suggesting that "they would be much more independent if they adopted the ordinary course of taking over the works of distribution and receiving a supply in bulk."[62] Each county council neighboring London, for example, could then make independent decisions about local water distribution and rates. The LCC deputation then protested that, while the County of London possessed 84 percent of the taxable value in the proposed board's area, the county council was given only ten seats on the Llandaff Commission's proposed thirty-member board.[63] Thus, London's water users would foot 84 percent of the bill for the companies' purchase while receiving only 33 percent of the representation on the new Metropolitan Water Board (this figure shrank to 20 percent in the government's final water board bill, 59 percent if municipal boroughs are included). Long responded that the government was obliged to remain faithful to the main features of the Llandaff Commission recommendations.[64]

The Progressives responded by making the question of London's water authority a cornerstone of their very successful 1901 election campaign. In it, they repeated the argument that the government's board was designed to undermine the LCC's power. Dickinson wrote in his campaign literature that the fact that the LCC would need to parcel out water to outside areas was not a legitimate justification for the government insisting on a water board. "This state of affairs is not particular to London, and in the case of many towns taking over the water supply, it has been provided for by the central municipality being made responsible for obtaining water for all requiring it and selling it in bulk to the neighboring authorities," he wrote. The need for sharing water was "a totally inadequate reason [for requiring] the establishment of a new body."[65] Thomas McKinnon Wood, the radical LCC water committee chair, wrote in his campaign pamphlet that "the Thames Conservancy is to have four members and the Lea Conservancy two [on the board]. . . . I do not know what purpose is served by introducing them in the Board except to assist to deprive it of municipal character."[66] A "Progressive Leaflet" declared that it was the aim of the Metropolitan Water Board that "the County Council . . . have less to do with Municipal functions, because it is the one authority which has stood up for London's rights and has dared to give expression to the aspirations of democracy."[67]

The Government Introduces Its Bill

Though the Progressives' 1901 election campaign was a great success, the Salisbury government pushed forward with its plans, officially introducing in January 1902 a bill to establish a water board. It was a ploy to decrease the LCC's influence, argued W. H. Dickinson in a newspaper interview a few days after the bill's introduction. He called the bill "a most barefaced piece of gerrymandering. London has shown at election after election that it does not want the Moderates to manage or mismanage the water question. But owing to the political lines on which the borough councils have been elected, the Water Trust will have a Moderate majority."[68] At a protest meeting two weeks later, John Burns, an MP and member of the Fabian Society, called the bill "a deliberate snubbing of London's chief body, a municipal injustice to London. . . . The Bill is part and parcel of a scheme to belittle the County Council and minimize its interest, with the view, if possible, of securing its ultimate abolition."[69] In House of Commons debate on giving the bill a second reading, Sidney Buxton, a London Liberal, said that "unless the . . . President of the Local Government Board . . . can give us some reasons he has . . . handed over these powers to the local bodies, giving them a dominating voice, I can only feel . . . that pressure has been put upon him by the London Conservative members sitting behind him, and who have always endeavoured to destroy the efficiency and the power of the London County Council."[70]

The LCC's majority also vocally opposed the government's water board on the grounds of sound governmental principle. "The proposed Water Board is unprecedented, irresponsible and without cohesion," stated the county council's resolution on the bill. In addition, "it is inexpedient and dangerous that a Board [be] constituted as proposed with no direct responsibility to the public."[71] The Progressives expressed grave concerns that the water supply, formerly solely subject to the decision making of company directors, now would be solely subject to the decision making of a board beyond easy reach of electors. W. H. Dickinson contributed an article to the *Morning Leader* in which he argued that the national government's plan was "contrary to municipal experience and Parliamentary precedent in regard to the water supply in our towns. . . . The supply of water should be in the hands of men directly elected by, and responsible to, the consumers."[72] Dickinson recognized that the government's stance was truly a deviation from the municipalization trend of the last half century. In House of Commons debate on the second reading of the bill, Buxton contended that "the right hon. gentleman in his bill seems to me to go as far away as he can from public control with this new body. . . . Such a proposal as this has never been attempted or suggested with regard to any of our great municipalities throughout the country."[73]

The Progressives and their supporters kept returning to the refrain that the water board was an attempt by the Conservative government to return, through indirect election, to the Metropolitan Board of Works— an effort, no doubt, to invoke images of corruption and indolence forever tied to the memory of that body.[74] Debating the second reading of the bill, H. H. Asquith, a Liberal MP and future prime minister, argued that "the late Metropolitan Board of Works . . . could not in the long run command the full confidence of the ratepayers of London. What is the reason? The moment you introduce this principle of indirect election . . . you open the door to the operation of influences and interests."[75] In LCC debate, E. A. Cornwall said, "Where is the Londoner of 1889 who, when the Metropolitan Board of Works died and was buried, thought that we should witness its resurrection in 1902? . . . It is not only a bad Bill, but it is a reactionary Bill, and . . . it is putting the hands of the clock back at least fifty years."[76]

The LCC's Progressive majority grew more outraged at the central government's plan for London's water supply when the Conservative majority amended the water board bill in the summer of 1902. The amended bill proposed that the Salisbury government select the water board's first paid chair and vice-chair. "The chairman and vice-chairman [of the new board] ought to be appointed by the Local Government Board," proposed Sir Frederick Dixon-Hartland, a Tory MP from Surrey and officer for the Primrose League, in the House of Commons. "I do not think that it ought to be left to chance—to a political chance," he said, "that the first man who wields great power and can make it work well, should be appointed in that way."[77] The amended bill also included the dictate that the first chair and vice-chair would serve two three-year terms in order to provide continuity during the first critical phase of the board's life. All major long-term policies would be decided by then, protested the LCC.[78] This proposal represented the closest possible control of the administration of the region's water supply by a small group at the center of the state. "As far as I am aware, it is totally unprecedented for Parliament to establish a board of men purporting to be representative and responsible to their respective local authorities, and at the same time to deprive them of the elementary right and the all-important duty of choosing their own Chairman," wrote W. H. Dickinson in the *Daily News*. "It is in reality introducing into this country the French system of Prefects instituted in the time of the Empire in order to curb the energies of the elected municipalities," he added.[79] Dickinson's picture of central control of vital natural resources differed from Sidney Webb's only in means and motives, not in degree.

Conservatives and their press allies, of course, offered rejoinders to the Progressives' opposition, the duty falling chiefly to Walter Long in the House of Commons. "If the London County Council feel themselves aggrieved because they have not by this time become the water authority for London it is themselves that they should blame rather than the Government or Parliament," he said when he introduced the bill, adding, "They did not secure the confidence of the community or this House."[80] And he argued that there were no grounds for the allegation that the board was designed with the purpose of undermining the LCC.[81] In a meeting of St. Pancras Conservatives discussing the bill, attendees agreed that the protests of the LCC were hollow because they were based on a mere power calculation. The LCC's was "a spurious indignation, founded on political considerations, because they knew that the Water Board, drawn as it was from the Borough Councils, would be a Conservative body and therefore out of harmony and sympathy with the Council."[82] The *City Press* agreed that the Progressives' opposition was based on "party animus alone." The *East London Observer* depicted the LCC's opposition to the government's solution in favor of their own as yet more evidence of the Progressives' great ambition. "But," concluded the newspaper, "their opposition is a forlorn hope."[83]

The Central Government's "Triumph" and the Purchase of the Companies

The bill was sent to a joint committee of both Houses on a straight partisan vote in March 1902.[84] There, the committee, though chaired by Lord Balfour of Burleigh, a Tory, voiced concerns over the unwieldy nature of the sixty-nine-member board. The committee went so far as to invite witnesses from provincial cities that had municipalized their water supplies to hear their testimony about water administration. The chair of the Liverpool Corporation water committee, the deputy town clerk of Glasgow, and others asserted that their small groups managed the water supplies of large areas successfully and efficiently.[85] The joint committee asked the Local Government Board if it would propose a smaller water board along the lines, at least, of that recommended by the Llandaff Commission. The Local Government Board refused (though, amazingly, Walter Long was open to the idea of a water board consisting only of representatives sent from the metropolitan boroughs).[86] The committee gave in to heavy government pressure and returned the bill to the House of Commons effectively unchanged.[87] There, the opposition of the LCC's allies attained some success, eliminating the new Metropolitan Water Board's government-appointed first chair and vice-chair and winning the right for the board itself to determine whether or not those officials were to be paid professionals; it seemed Long was willing to concede the item because he could rely on the inevitable Conservative majority on the board to select a suitable executive.[88] Save for that one change, an important one in the eyes of the Progressives, the bill remained "beyond redemption," in the words of Henry Campbell-Bannerman, Liberal MP and future prime minister.[89] In late 1902, "the Bill was triumphantly placed on the Statute Book," wrote Walter Long, as the government received its Metropolis Water Act.[90] The LCC and Fabians declared the act "retrograde," "discredited," and "plutocratic."[91]

The Metropolitan Water Board came into being in April 1903. The LCC sent Progressive and Moderate representatives in the same proportion as sat on the council—ten and four, respectively.[92] It certainly appeared that the LCC's interest in water supply went beyond a mere political power calculation to political principle since, though they had little hope of influencing a politically composed board dominated by their party rivals, some of the LCC's leading Progressive figures chose to serve in the role. Among others, the one-time council chair and former water committee chair W. H. Dickinson, the recent water committee chair and Progressive leader T. McKinnon Wood, the LCC deputy chair Henry Clarke, the longtime committee member T. H. W. Idris, and John Burns were vocal participants in the new water board's early meetings.[93] Sidney Webb became a member in 1904.

In the first few meetings of the Metropolitan Water Board, held in the Privy Council's meeting room, the LCC's representatives argued that some "municipal character" could be salvaged for the board. In the words of John Burns, it should vote to choose a chair and vice-chair from among their number, "from the elected sphere."[94] The MWB could otherwise easily become a closed-off bureaucracy, a fiefdom of experts if professional administrators were hired. "If we are going to set up a paid Chairman of this Board," warned W. H. Dickinson, "we depart, I venture to say, seriously from the whole principle that has guided municipal administration, not only in London, but all over the country."[95] But the Progressives' desire to influence the new board appears to have been another forlorn hope. The question of a paid executive, it seems, was settled before the LCC's leaders even spoke. A committee of the board's leading Conservatives met prior to the MWB's inauguration and agreed among themselves on its organizational structure and the duties of each official, made arrangements with the Local Government Board for short-term funds, and entered into other agreements.[96] The board's leadership would have a professional character, with a former water company director elected as its head and compensated on a paid basis.[97]

The Metropolitan Water Board was to take over all of the water companies' operations in June 1903 and in the meantime had to take part in arbitrations in which a total of tens of millions of pounds of public money was at stake. The companies' claims for compensation began arriving in the fall, eventually totaling around £60 million.[98] The figure represented the total of the companies' paid-up capital and capital expenditures, their net profits for 1903-4, the value of their stock and cost of their back dividends, the costs of shutting down operations, and their debts.[99] The cost to taxpayers would have been much higher if the LCC had not been pursuing municipalization throughout the decade; it had succeeded in getting clauses inserted in a number of parliamentary acts authorizing water companies to raise funds for new works, and those clauses effectively barred the companies from claiming reimbursement for the revenue in the event of public purchase—that is, the LCC barred the companies from borrowing money from investors and then investing it in works for the purpose of getting reimbursed by taxpayers. The 1902 water board act also barred the arbitrators from following the Lands Clauses Consolidation Act of 1845, which required government bodies to pay a premium, usually

10 percent, on private property they expropriated. The LCC had argued for this principle throughout its decade-long effort to acquire the companies. Forcing an industry to sell its operations, it had contended, was fundamentally different than compelling a homeowner or property owner to abandon land.

The arbitrators settled on a price—well below the £60 million sum the companies demanded—by estimating how well the companies' stock could have been expected to perform in the future, based on their past performance, had the companies not been purchased.[100] Shareholders were compensated in Metropolitan Water Board stock, which paid guaranteed dividends of 3 percent annually, a total of £30.6 million. On top of this, the MWB was burdened with £450,000 in various expenses, including compensation to company executives, and was saddled with paying £1.5 million in interest annually.[101] The purchase price would have been lower if the London County Council had succeeded in buying the companies. The LCC had always insisted that the cost to buy out the companies must reflect the fact that they were approaching a dead end in terms of their water sources—that their success was not guaranteed for fifteen years, let alone perpetuity. The arbitrators refused to consider the companies' shortcomings and for the purpose of determining a purchase price assumed that their profits were perfectly sustainable.[102] And by mere virtue of delaying the moment of purchase, the Salisbury government greatly increased the eventual cost to London taxpayers; with each passing year, the capital expenditure of the companies increased while every five years London property values were reassessed, with the result that the taxable value of London property greatly increased.[103]

Having surveyed the previous generation's efforts to municipalize London's water supply, the prominent Progressive W. H. Dickinson wrote in 1902 that "it seems an extraordinary position of affairs that His Majesty's Government should feel it their duty to force upon London a new system of administration."[104] In the history of the water municipalization movement, the degree of the central government's involvement in the London case, the methods of its intervention, and the motives for its actions were extraordinary. It is not that Parliament had completely distanced itself from cities' municipalization and waterworks construction efforts in the past, but its involvement then had been limited and usually permissive.[105] Town councils frequently brought local private bills to Parliament only after they had made an agreement with the water companies they aimed to purchase. When they had to purchase companies by compulsion, Parliament usually simply verified that companies and their stockholders were well treated—which they always were.[106] In 1878, the House of Lords blocked Manchester's bill for transforming Thirlmere in the Lake District into a reservoir on the stated grounds of technical noncompliance but more likely on the grounds that it sympathized with opponents who argued that the lake would be disfigured.[107] The next year, Manchester's act permitting its scheme went through. In 1866, Parliament had been dissatisfied with Huddersfield's bill for new waterworks construction because of concerns about the safety of an embanked reservoir in proximity to the population. Parliament sent it back to the drafting board.[108]

In the case of London, the central government intervened first by halting the London county government's efforts to purchase the cities' eight water companies in the mid-1890s. It continued involving itself by blocking the county council's efforts to introduce a new water supply into the metropolis in the second half of the decade. At the end of the decade, the Salisbury government instituted a royal commission to consider solutions to the water question, thus stalling London's water initiatives until the commission reported near the eve of 1900. Then the government declared an end to the controversy; it called for a new government for water based on the fundamental idea expounded by its commission but with a constitution of its own elaborate specifications to empower its creations—the metropolitan boroughs.

The central government's actions were an effort to achieve a social outcome through the way it administered the environment, just as were Bradford's, Birmingham's, and Manchester's actions. The efforts of those towns had their roots in environmental crises specific to expanding industrial towns and were fueled by the ideals of individuals who offered a vision of the modern city. In London's case, the actions of the national government were in response to an environmental crisis that laid bare commercial water suppliers' shortfalls in the midst of a population boom (an increase of four hundred thousand between 1881 and 1891) and unexpected weather conditions.[109] Those national leaders reforming London's water operations were also driven by ideals, by an alternative vision of modernity—there was hardly a single conception. Those dominating the central government were apprehensive that, in London, a major component of modernization would be collectivization. They did not want the modernization of London to mean the centralization of power in the hands of a body "almost a competitor with the House of Commons," led by activist, doctrinaire officials with the sorts of aims outlined in *The London Programme*.[110] So the state invented an administrative body that reflected the outcome it desired for an environmental resource. The national government wanted the control of water out of the hands of a central body, so it put water largely in the hands of a fractured collection of local authorities—the metropolitan boroughs. It wanted to keep water (and its latent power to alter society) from serving a political ideology, so it tried to put water in the hands of professional administrators. It wanted to keep water from empowering a public body that the central state mistrusted, so it invented a disembodied organization. The central government's vision of the future was less vibrantly portrayed than that of the Fabians—it was more reactive than inventive—but preserving the power of water was key to realizing it.

Conclusion

This story opens at a moment of change, on the eve of a period when the rapidly expanding urban environment forced a transformation in the nature of British urban government. In the growing industrial towns of the first half of the nineteenth century, industrial waste and the concentration of large populations created acute environmental pressures. In this period, water, a primary necessity for human life, was both hard to obtain and hard to dispose of after use. It was difficult for townspeople to eliminate spent water from their habitations because of inadequate drainage systems and difficult for them to secure clean water in an environment glutted with lingering waste and lacking adequate water supply systems. Existing water companies were notoriously poor at supplying water of good quality and on a dependable basis to a majority of the population. Solving these problems demanded constitutional changes in the urban governments that had come into being—though without many specified powers—through the Municipal Corporations Act of 1835. Beginning in that year, councils applied to Parliament to expand their power and financial strength so that they could address water crises and avert disasters such as epidemics. Councils proceeded by buying commercial water operations, expanding their sources of supply, and building drainage systems. The implication of this development for historians is that contingent environmental factors had the power to generate political modifications. The significance of this stage in terms of this narrative is that, from this moment, supplying water to the populace was recognized as a primary duty of new local governments.

Soon, a segment of the urban elite began to see that water supply held greater importance than just practical considerations, such as preventing disease outbreaks. These elite leaders began to imagine their cities flush with water, with fountains, public baths, and aqueducts that would deliver pristine water from remote hills. They pictured their community's productivity and population freed from any limits. And they began to envision the poor, long imagined to be unable to clean themselves and their abodes, raised to virtue by a new ease of access to water. These civic leaders coupled environmental pollution with moral pollution, and they took it upon themselves to deliver water to every home, no matter how lowly. They saw themselves as part of a new breed of moral governors. This development marks a turning point in the history of water, a phase when water came to stand for much more than itself—it would become an instrumental force for realizing a new kind of civic society.

To fulfill this vision, the majority of cities that took over local water companies quickly augmented the existing systems by building new—often massive—waterworks. These projects further concentrated power in the hands of councils because, to develop new sources of supply, cities appropriated land outside of their limits and had to gain parliamentary sanction to issue stock to finance the purchase of land and pay for construction. A new prevailing model of waterworks, engineers claimed, satisfied all the requirements of the new water supply ideal—it provided pristine, abundant water under constant pressure and available to any location in the city twenty-four hours a day. With that, the new environmental projects and works that delivered this health- and morality-boosting element took on symbolic importance. Modernizing waterworks represented physical environmental change in the urban hinterland that was to have an enlightening effect on the distant city. This profound shift offers a challenge to the long-standing conventional wisdom that modernity tended to proceed at the command of the central state, which executed large-scale projects that generated environmental change or rationalization for the purpose of realizing social change. This case, as well as alternative choices for administering water supply, offers a different mode of modernity and mechanisms of modernization.

Britain's capital faced the same environmental pressures as other industrial cities or rather even more, being one of the largest cities in the world. From early in the nineteenth century, London's would-be reformers fulminated about its water problems, but as the decades passed and provincial towns changed their environments, London accomplished little. Only at the point of an epidemiological gun in the form of looming cholera did London construct a sewage system in the middle of the century. This achievement fulfilled only part of the goal reformers envisioned; in preceding years, they had called for a local government with a strong enough constitution to obtain parliamentary approval of new financial powers allowing them to buy out the water companies and to augment existing water supplies. London's only metropolis-wide authority, the Metropolitan Board of Works, lacked the status that would enable it to take on the water companies. It had had enough power to undertake the Main Drainage scheme and the Thames Embankment—Joseph Bazalgette's works that expelled the metropolis's wastewater and protected Londoners from cholera—but not to overcome the powerful interests whose livelihoods depended on the sale of water.

Near the end of the nineteenth century, London finally did gain an authoritative local government as the natural outgrowth of the Local Government Act of 1888, which enabled the formation of county councils throughout England and Wales. The new London County Council replaced the Metropolitan Board of Works in 1889, and there was widespread anticipation among water activists that now, finally, London had a suitable governing body to join the water municipalization movement—if half a century late. According to many, operating the community's water supply was simply what the modern municipal government did. There could be no more important responsibility. The members of the LCC's first session, at least, assumed so, and—with agreement in the body that largely crossed all political and ideological lines—the council immediately proceeded along the same well-established course that other cities had taken to become the exclusive supplier of water.

By the council's second and third three-year sessions, however, the LCC had a much more radical complexion. Its Progressive majority had come under the influence of a platform laid out by the Fabian Society —a small but articulate group of socialists who sought, by gradualist means, to change society's basic economic organization. It wished to make a moral life possible for all members of society instead of seeking to change only the behavior of those of the lowest station. Only by living in a just society that freed all its members from participating in a system that perpetuated inequality, the members of the society believed, could all citizens live a moral life. The Fabians sought to change the flow of wealth in London as if they were trying to reverse the flow of a river. As things stood, the income of the city's average worker flowed in one direction—up to a relatively small population of individuals who owned most of London's land and resources; the Fabians wanted to redirect the flow of the value of those resources outward to all citizens equally. In the Fabians' vision, a group of democratically elected overseers at the center of the city would govern the supply and distribution of gas, water, and transportation; they would direct a vast labor force, operate the port, maintain the Thames, and so on. Income from municipal industry and utilities would flow to the center and then back out again on the basis of equalizing living conditions, improving the efficient and productive operation of the city, all with the purpose of bringing more sectors of the municipality under communal control. For the leaders of the Fabian Society, ordering the environment was the key to social change, just as it had been for leaders in provincial cities in the past, but the Fabians were driven by a much more far-reaching purpose than that of urban reformers who had gone before. Water began to stand for something drastically different—not social amelioration, but fundamental transformation.

In the eyes of those watching from the center of the national government—the Conservative leaders of the Salisbury government—the water municipalization movement looked very different from this point onward. These leaders believed that, in the hands of the Progressives and their Fabian allies, water reform had come to represent a dangerous slide toward collectivization. The Conservatives envisioned a contrary future, a modern London in which state enterprise would be kept to a bare minimum, the centralizing tendencies of the LCC rejected in favor of more localized control, and unpredictable democracy tempered with oligarchy. In their view, gas, water, locomotion, labor, housing, the Thames—the chief elements of the urban ecosystem—should not be in the hands of the Fabian and Progressive puppet masters to manipulate for their own ends but in a group of private and, frankly, Conservative hands. Now, conflicting visions of modernity clashed over water. This development heralded a break from the preceding history of water reform. Parliament had consistently granted town councils' requests to municipalize their water companies in the past; the central government had kept out of the issue. Now it intervened. The LCC promoted bills for purchasing the water companies, and the Conservative majority in Parliament, by refusing to grant the council an act that would force the companies to enter into binding arbitration to settle the terms of their sale, blocked the bills. The LCC promoted a modernistic waterworks scheme, but recognizing that modern waterworks could fix a policy in place, the Conservatives threw their weight behind a rival scheme. Things stood at an impasse.

In the mid-1890s, another entity—extreme weather conditions—forcefully entered the conflict between the political rivals. A lack of rainfall and high temperatures caused water reservoir levels to drop dramatically, and consumers found their taps dry except for brief periods during the day. Londoners became justifiably alarmed. Conservatives had supported private enterprise—that is, they had supported the commercial water suppliers of London—in their resistance against the LCC's takeover attempts, but drought had revealed the companies' serious inability to deal with shortfalls in supply. With public agitation adding to their wariness, the leaders of the Conservative central government had to act. The extreme weather event laid bare the operation of the urban ecosystem. Water in the city existed in a close relationship with political structures, economic systems, human ideals, and thousands or millions of human bodies that could not go for even a day without it; when water defied expectations in a way that endangered the population and elicited public outcry, something had to change on the political side of things.

The solution proposed by the central government demonstrated that a system of environmental administration could be designed to fix a political philosophy in place and prohibit an alternative from gaining ground. The Conservatives decided to amalgamate the water companies and to have their operations run on a nonprofit basis. For them, the goal of a new, public water operator must be to provide a sufficient supply of water of acceptable quality as inexpensively as possible. Although the Conservatives' plan lacked the underlying goal of improving society that civic water reformers of the past had pursued, the Conservative plan was in its essentials somewhat similar to the initiatives in provincial cities. The government's objectives were thoroughly remote, however, from the goals the Fabians and Progressives had sought via improving water supply. The national government created a water board with a constitution that precluded the organization from serving the aims of the LCC, decentralized influence within the board, empowered London's second-tier authorities to present a challenge to the LCC, and gave less influence to democracy and elected officials and more to professionals and traditional oligarchies. This outcome, with Parliament stepping in to place a "water government" over the London region—usurping London's elected authority—appears drastic unless the political consequences that rested on the water issue are understood.

Parliament's intervention in London water matters demonstrates the significance it placed not only on the issue but also on London itself. Although all during the century reformers like the Westminster vestryman James Beal and Joseph Firth had charged that Parliament was negligent in its handling of the London government issue, in actuality Parliament took great interest in London's affairs, especially when public health or Parliament's power was threatened. True, as the reformers had complained, Parliament was not eager to give London a strong central authority. In the central government's view, an authority for all of London had the potential to absorb inordinate amounts of wealth and power and even had the potential to dictate the course of the nation virtually independent of Parliament. Thus, the Salisbury government had trepidations about granting London a county council, so when it finally did create the LCC, it withheld some fundamental powers from the new body. Although Parliament handled the question of London government cautiously, the history of London water supply suggests that Parliament was mindful of London's administration. The national government could indeed intervene, and quite boldly. The construction of the Thames Embankment and the related Main Drainage system—undertaken by a metropolitan body established by Parliament for that very purpose—was a project on an unprecedented scale, an environmental reform of a magnitude not seen since in London. The second great intervention of the period—the creation of the Metropolitan Water Board—constituted the transfer of tens of millions of pounds of capital from private to public hands and the total reorganization of the administration of London's water supply. Parliament proved that it watched closely the direction of London—more closely, in the case of water, than it observed provincial cities. Parliament allowed the water modernization movement to proceed of its own volition in the provinces, but in London, where more (more power, in particular) was at stake, it did not. When pressed, those meeting in Westminster took steps to see that water was administered in a manner that achieved Parliament's goals for the wider metropolis surrounding the core.

Governments took on the task of supplying water to serve their own ends, but this case shows that water could also affect the fate of political groups. Losing the contest over water to the central government meant that the Progressive-dominated London County Council suffered a significant loss of power and responsibility. Supplying water would have been one of the council's principal duties. It represented significant revenue—and potentially a profit source, though this is uncertain—and in one sense made the council a ubiquitous presence in the very households of all Londoners. The LCC's loss of the authority to supply water meant the loss of one of its inaugural initiatives, a cornerstone of the Progressives' "London Programme," and of a reform that would likely have come much more cheaply than its other projects did.[1] The council had expended thousands of pounds on its effort to take control of water, from research expenses to drafting bills for Parliament, as well as hundreds of committee hours and thousands of employee hours. The engineering department alone had been consumed by the initiative for years. The council's reputation was invested in its vow to purchase the companies and in its Welsh scheme, and the 1901 election showed that electors counted on the LCC to act and succeed. However, London's second-tier local authorities (the metropolitan borough councils) had a higher combined representation on the new Metropolitan Water Board.

When the national government created the borough councils, they gained significant power at the expense of the London County Council and the Progressives that served on it. Progressive county councilors, the LCC's press allies, and allies in Parliament made it clear that, though the water companies had been eliminated, the creation of the new Metropolitan Water Board did not mean a victory for the LCC. Instead, they described it as a blow to the council's dignity and a general defeat of the principles of modern municipal reform.

The Liberal MP Sidney Buxton predicted, during debate prior to passage of the Metropolis Water Act of 1902, that the proposed Metropolitan Water Board would "be inefficient, uneconomical, and nonrepresentative."[2] He had grounds. The board, in the first place, was far larger than necessary—far larger than the town council committees that administered large water systems elsewhere in Britain. The financial terms that the government offered the companies, too, were far more generous than those that the LCC had proposed. Finally, as Buxton implied, the water board was also far less directly representative than was the LCC. The largest bloc of board members comprised nominees sent—one each—from the metropolitan borough councils. These delegates, in turn, had been elected by one small ward within each borough. Some representatives sent to the water board from the wide variety of its constituent bodies had never won any election at all. Had the LCC become the water authority, the Progressives maintained, London's electors could have directly controlled their water managers, rewarding or punishing representatives' performance every three years at the polls—even drastically overhauling the profile of the council at a stroke, if necessary.

Still, the MWB enjoyed a relatively successful career, democratic or not. Problems of quality and sufficiency did not vanish with the demise of the water companies. Complaints over turbid water and shortages in limited areas continued to occur.[3] But these were less the fault of the board's oligarchical or undemocratic nature than they were the result of the difficulty in expanding the water system's infrastructure at a pace to keep up with demand.[4] Indeed, the water board spent its lifetime struggling to keep up with demand as the population reached a peak of 8.6 million on the eve of World War II. And, in this situation, it cannot fairly be said that the MWB failed London. At a steady rate, the board constructed massive reservoirs to hold more and more Thames water when the river swelled during the winter months, until the banks of the Thames and its tributaries west of the metropolis were crowded with human-made lakes encased in brick. The Progressives' predictions that the Thames would fail London went unfulfilled, and so did fears that growing populations in the Thames valley would pollute London's water supply and lead to epidemic. The water board safeguarded Londoners by introducing chlorine into its supplies at the close of World War I.[5]

The Metropolitan Water Board expired in 1973 when Parliament created ten main water authorities for England and Wales in a new water act. In 1989, Margaret Thatcher's government returned London's water supply to private hands, giving the closely monitored Thames Water Company a state-guaranteed monopoly for twenty-five years. Three years earlier, her government had eliminated the Greater London Council, the successor of the London County Council, and transferred its powers to the metropolitan borough councils. Water was then back in private hands, the last vestiges of the LCC were no more, and the borough councils were further empowered. It was an outcome of which Lord Salisbury would have approved.

London receives its drinking water from the river Thames and regional underground sources. That is putting it very simply, because London actually receives its water by way of a large number of environmental, technical, legislative, and corporate intermediaries. Taking this further, London receives its water amid myriad debates, including wrangling over water-quality science, over standards of regulation, and over the performance of its for-profit water utility. And the most powerful intermediary, of course, is the weather. All of these factors and more form a network, or regime, or ecosystem amid which and by which water supplies are collected and distributed to consumers.[6]

If there is stress in one area of the regime or branch of the ecosystem—the climatological area, for example—the success of the water supply is threatened. And London's water regime is threatened today. Greater variability in rainfall and more dry, hot summers are the main stresses, and they are revealing vulnerabilities in other parts of the water supply ecosystem.[7] When there is a lack of rainfall and water utilities impose water use limits, consumers voice frustration, pointing out that the water utilities enjoy great profits even while they deny consumers their product, and the gardening industry complains of lost sales.[8] When there is stress among the technological components of the ecosystem, such as leaky pipes or insufficient storage, consumers are further angered by company profits, watering bans, and rate increases.[9] Consumers, elected representatives, and regulators then question which part of the ecosystem is failing: is it the technological, the regulatory, or the management aspect of the utility? And consumers and political figures have called for overhauling the regime, beginning with the sources of supply and ending with the company delivering it.[10] Put in other words, recurrent stresses in the environmental aspect of the ecosystem have led to stresses in the technical, which naturally affect the social, which has resulted in rumblings in the political branch. It is possible that this form of the water supply ecosystem is unsustainable, that components of it will collapse and be replaced, and that a new system will emerge, assembled through a mix of means—political, technological, accidental—by a mix of players and factors.

Now, as stresses delineate the radiuses of London's water ecosystem distinctly, is a propitious time to examine the prehistory of how this network came into being. It is a time to look past the origins of Thames Water, past Margaret Thatcher's sale of the regional water authorities that created it, past the creation of the RWAs in the first place, even past the creation of the first large water authority in Britain, the Metropolitan Water Board. To look back at a time before there was any water company or water authority for London reveals the deep roots of discourses about appropriate technologies, appropriate water standards, appropriate water supply governance, and even appropriate urban governance, broadly speaking.

Then, as now, a range of actors debated the exact nature of the problems of water in the city under stress. Then, as now, potential solutions vied for supremacy amid the network. And just as tomorrow's solutions to scarcity and other challenges will shape society in ways seen and unforeseen, so did nineteenth-century ones powerfully shape British urban society and politics.

Notes

All newspapers cited were published in London unless otherwise indicated.

Introduction

1. James Smith, *Report on the State of the City of York and Other Towns* (London: W. Clowes and Sons, 1845), 8, 13.

2. William Kay, *Health of Towns Commission, Report on the Sanatory [sic] Condition of Bristol and Clifton* (n.p., 1844), 2, 10, 32.

3. James Ranald Martin, *Report on the State of Nottingham and Other Towns* (London: W. Clowes and Sons, 1845), 36.

4. Sally Sheard, "Nineteenth Century Public Health: A Study of Liverpool, Belfast and Glasgow" (DPhil thesis, University of Liverpool, 1993), 98; J. A. Hassan, "The Growth and Impact of the British Water Industry in the Nineteenth Century," *Economic History Review*, 2nd ser., 38, no. 4 (1985): 534.

5. Hassan, "Growth and Impact of the British Water Industry," 538.

6. *Bradford Observer*, 9 October 1851, 4.

7. [J. Hill Burton,] "Sanitary Reform," *Edinburgh Review* 91 (January 1850): 220. The article was not signed; the attribution to Burton comes from the Wellesley Index to Victorian Periodicals.

8. Harry Chester, "The Food of the People, Part II," *Macmillan's Magazine* 19 (November 1868): 19; Brighton Health Congress, *Transactions of the Brighton Health Congress* (London: Marlborough, 1881), xiii.

9. A misperception persists in the literature that the creation of the London County Council (LCC) in the late 1880s meant that London had finally overcome the last obstacle to administering its own water supply. As will be made clear, a later entity, created specifically to limit the LCC's influence, became the eventual long-term water authority. Besides, the Metropolitan Board of

Works formed in the mid-1850s had had the desire and potential to be London's sole water provider far earlier. Debora Spar and Krzysztof Bebenek, "To the Tap: Public versus Private Water Provision at the Turn of the Twentieth Century," *Business History Review* 83 (winter 2009): 688.

10. My indebtedness to Christopher Hamlin, *Public Health and Social Justice in the Age of Chadwick, Britain, 1800–1854* (Cambridge: Cambridge University Press, 1998), 9 and elsewhere, is obvious. For earlier, simplistic accounts, see George Rosen, *A History of Public Health*, rev. ed. (Baltimore: Johns Hopkins University Press, 1993), 230, 240, 259, and elsewhere; Dorothy Porter, "Introduction," in *The History of Public Health and the Modern State,* ed. Dorothy Porter (Amsterdam: Editions Rodopi, 1994), 1–44; and Raymond Goldsteen, *Introduction to Public Health* (New York: Springer, 2011), 60.

11. Recent histories have tended to focus on water drainage rather than water supply. See Hamlin, *Public Health and Social Justice in the Age of Chadwick;* J. V. Pickstone, "Dearth, Dirt and Fever Epidemics: Rewriting the History of British Public Health, 1780–1850," in *Epidemics and Ideas: Essays on the Historical Perception of Pestilence,* ed. Terence Ranger and Paul Slack (Cambridge: Cambridge University Press, 1992), 125–48; Patrick Joyce, *The Rule of Freedom: Liberalism and the Modern City* (London: Verso, 2003), esp. 62–75; Thomas Crook, "Norms, Forms and Bodies: Public Health, Liberalism and the Victorian City, 1830–1900" (PhD diss., University of Manchester, 2004). Dealing less explicitly with water is Christopher Otter, "Making Liberalism Durable: Vision and Civility in the Late Victorian City, 1870–1900," *Social History* 27 (2002): 1–13. Focusing more squarely on water supply is John Broich, "Engineering the Empire: British Water Supply Systems and Colonial Societies, 1850–1900," *Journal of British Studies* 46 (April 2007): 346–65.

Chapter 1. Water and the Making of the Modern British City

1. Anne Hardy, "Water and the Search for Public Health in London in the Eighteenth and Nineteenth Centuries," *Medical History* 28 (1984): 251; Walter M. Stern, "Water Supply in Britain: The Development of a Public Service," *Royal Sanitary Institute Journal* 74 (October 1954): 998.

2. John Frederic La Trobe Bateman, *History and Description of the Manchester Waterworks* (Manchester: T. J. Day, 1884), 232.

3. John Burnet, *History of the Water Supply of Glasgow* (Glasgow: Bell and Bain, 1869), 2.

4. Jack Loudan, *In Search of Water, Being a History of the Belfast Water Supply* (Belfast: Mullan and Son, 1940), 3.

5. *History of the Water Supply of Newcastle-upon-Tyne* (Newcastle: Newcastle Chronicle, 1851), 7.

6. John Graham-Leigh, *London's Water Wars: The Competition for London's Water Supply in the Nineteenth Century* (London: Francis Boutle, 2000), 9.

7. A. L. Dakyns, "The Water Supply of English Towns in 1846," *Manchester School* 2 (1931): 20.

8. J. A. Hassan, "The Growth and Impact of the British Water Industry in the Nineteenth Century," *Economic History Review*, 2nd ser., 38 (1985): 532.

9. G. Kitson Clark, *An Expanding Society, Britain 1830–1900* (Cambridge: Cambridge University Press, 1967), 4. The rate of growth continued; Britain's population increased 120 percent between 1841 and 1911. C. H. Lee, "Regional Growth and Structural Change in Victorian Britain," *Economic Historical Review*, new ser., 34 (August 1981): 441.

10. Roy Porter, *London: A Social History* (Cambridge, MA: Harvard University Press, 1994), 205–38; *Leicester Water Undertaking, 1847–1874* (Leicester: Leicester Water Department, 1974), 58–59; Peter Ellerton Russell, "John Frederic La Trobe-Bateman, F.R.S., Water Engineer 1810–1889" (MS thesis, University of Manchester, 1980), 2; Asa Briggs, *Victorian Cities* (New York: Harper and Row, 1965), 88–89. Northampton doubled in population between 1801 and 1831, Huddersfield grew by 30 percent in every decade of the first half of the century, Brighton's population doubled nine times, and Cheltenham's doubled eleven times. Ibid., 364–65.

11. Briggs, *Victorian Cities,* 59.

12. David Roberts, *Victorian Origins of the British Welfare State* (New Haven: Yale University Press, 1960), 2.

13. For the debate on rising population and marriage and fertility in cities, see Kenneth Morgan, *The Birth of Industrial Britain* (London: Longman, 1999), 8–13; on dietary improvement, see Thomas McKeown, R. G. Brown, and R. G. Record, "An Interpretation of the Modern Rise of Population in Europe," *Population Studies* 26 (November 1972): 347.

14. Henry Jephson, *The Sanitary Evolution of London* (London: T. Fisher Unwin, 1907), 155.

15. Morgan, *Birth of Industrial Britain,* 33–38.

16. Burnet, *History of the Water Supply of Glasgow,* 2–3.

17. Hassan, "Growth and Impact of the British Water Industry," 533.

18. Dakyns, "Water Supply of English Towns," 20.

19. Hassan, "Growth and Impact of the British Water Industry," 533–34; Dakyns, "Water Supply of English Towns," 21.

20. Burnet, *History of the Water Supply of Glasgow,* 3–4.

21. Arthur Redford, *The History of Local Government in Manchester,* 3 vols. (London: Longmans, Green, 1940), 1:171–74; *Leicester Water Undertaking,* 59; T. W. Woodhead, *History of the Huddersfield Water Supplies* (Huddersfield: Wheatley, Dyson & Son, 1939), 48; *History of the Water Supply of Newcastle,* 30–32. See also Hassan, "Growth and Impact of the British Water Industry," 534.

22. Hardy, "Water and the Search for Public Health," 252.

23. Stern, "Water Supply in Britain," 999.

24. Elie Halévy, "Before 1835," in *A Century of Municipal Progress, 1835–1935,* ed. Harold Laski, W. Ivor Jennings, and William A. Robson (London: George Allen and Unwin, 1935), 19.

25. Derek Fraser, ed., *Municipal Reform and the Industrial City* (New York: St. Martin's Press, 1982), 4; Halévy, "Before 1835," 19.

26. Fraser, *Municipal Reform,* 4.

27. Derek Fraser, *Power and Authority in the Victorian City* (New York: St. Martin's Press, 1979), 4.

28. The inclusion of a large number of aldermen—presumably selected from among the great men of the town—was, according to Fraser, a concession to the House of Lords, which naturally objected to the landed interests' loss of the corporations. Apparently, members of the House of Lords expected the aldermen to be their kind of men. Fraser, *Power and Authority,* 10.

29. Fraser, *Municipal Reform,* 4–5.

30. Ibid., 5.

31. Ibid., 6.

32. See Fraser, *Power and Authority,* 164.

33. Fraser, *Municipal Reform,* 8.

34. Chadwick was secretary to the Poor Law Commission, where reports from Poor Law guardians and workhouse doctors constantly crossed his desk and provided a picture of death and disease. See R. A. Lewis, *Edwin Chadwick and the Public Health Movement, 1832–1854* (London: Longmans, Green, 1952), 33–34.

35. Chadwick also solicited information on area conditions from more than two thousand physicians and local officials throughout Britain. M. W. Flinn, introduction to Edwin Chadwick, *Report on the Sanitary Condition of the Labouring Population of Great Britain* (London: Clowes and Sons, 1842; repr., Edinburgh: University Press, 1965), 47–51.

36. Chadwick would play a leading role in devising the new Poor Law, which reorganized the provision of poverty relief, and in influencing the creation of the unprecedented central Board of Health in 1848. In his rise to influence, he also invited extraordinary resentment for his dogmatism and ambition. See, generally, Anthony Brundage, *England's "Prussian Minister": Edwin Chadwick and the Politics of Government Growth, 1832–1854* (University Park: Pennsylvania State University Press, 1988).

37. Lewis, *Edwin Chadwick and the Public Health Movement,* 38.

38. Edwin Chadwick, *Report on the Sanitary Condition of the Labouring Population of Great Britain* (London: Clowes and Sons, 1842), 120, 196–97. For the prevalence of beer drinking, see ibid., 62.

39. Chadwick, *Report on the Sanitary Condition of the Labouring Population,* 63.

40. Chadwick stated that the harmful effects of sewage were halted by

"putting the solid manure into water to arrest the gases." Edwin Chadwick, "Discussion on 'The Utilisation of the Sewage of Towns,'" *Journal of the Society of Arts* 212 (12 December 1856): 49. On the prevalence of miasma theory, see Dale H. Porter, *The Thames Embankment: Environment, Technology, and Society in Victorian London* (Akron: University of Akron Press, 1998), 55.

41. Chadwick, *Report on the Sanitary Condition of the Labouring Population*, 341. Chadwick focused on flushing the air, too, but this would be accomplished mainly by rinsing and submerging human and industrial wastes.

42. Chadwick's superiors at the Poor Law Commission did not want to bear any responsibility for such a sensational tract, bound to shock readers and give offense to local government bodies such as the sewerage boards that Chadwick condemned. Lewis, *Edwin Chadwick and the Public Health Movement*, 40.

43. Roberts, *Victorian Origins*, 72.

44. Such was the case for the Leicester Waterworks Company and Manchester and Salford Waterworks Companies, for example, which relied on wells and brooks, respectively, that had been long used. Scott Edward Roney, "Trial and Error in the Pursuit of Public Health: Leicester, 1849–1891" (PhD diss., University of Tennessee, Knoxville, 2002), 78–79; Redford, *History of Local Government in Manchester*, 1:174.

45. Lyon Playfair, *Report on the State of Large Towns in Lancashire* (London: W. Clowes and Sons, 1845), 57. Many of the commissioners' reports for individual cities and industrial regions were published as separate pamphlets.

46. *Second Report of the Commissioners for Inquiring into the State of Large Towns and Populous Districts* (London: W. Clowes, 1845), 48.

47. Bernard Rudden, *The New River: A Legal History* (Oxford: Clarendon Press, 1985), 164.

48. *Second Report of the Commissioners*, 51.

49. Ibid., 48; T. Laycock, *Report on the State of the City of York* (London: W. Clowes, 1844), 2.

50. *Second Report of the Commissioners*, 47.

51. Roberts, *Victorian Origins*, 73.

52. Liverpool correspondent quoted in *Second Report of the Commissioners*, 292. Chadwick had expressed this belief as well. "Improvement creates expense, which is felt in diminution of the dividends of stockholders," Chadwick explained, which "creates strong interest against all improvements in the quality of the supplies of water." Chadwick, *Report on the Sanitary Condition of the Labouring Population*, 144.

53. *Second Report of the Commissioners*, 50. Chadwick had arrived at the same conclusion. "Arrangements should be made for all houses to be supplied with good water," he wrote in his 1842 report, adding that "for this purpose . . . power should be vested in the most eligible local administrative body." Chadwick, *Report on the Sanitary Condition of the Labouring Population*, 150.

54. *Second Report of the Commissioners,* 2.

55. Ibid., 50.

56. See, for example, the discussion of opposition to a strong central Board of Health in Roberts, *Victorian Origins,* 74–77. Britain, of course, was not wholly in the grasp of the laissez-faire ideal. For a convincing argument that there was a trend of government intervention alongside the rise in the ideal of nonintervention in issues involving private property, see J. Bartlet Brebner, "Laissez Faire and State Intervention in Nineteenth-Century Britain," *Journal of Economic History* 8, supplement S1 (January 1948): 59–73. Also see Harold Perkin, "Individualism versus Collectivism in Nineteenth-Century Britain: A False Antithesis," *Journal of British Studies* 17 (autumn 1977): 105–18.

57. The act also created a central Board of Health, on which Chadwick served, but it soon was dissolved, a victim of opponents of centralization.

58. Sally Sheard, "Nineteenth Century Public Health: A Study of Liverpool, Belfast and Glasgow" (DPhil thesis, University of Liverpool, 1993), 99.

59. *Second Report of the Commissioners,* 2. Arthur Redford believes the reports brought things to a head, as well. See Redford, *History of Local Government in Manchester,* 1:177.

60. See, for examples, Playfair, *Report on the State of Large Towns in Lancashire;* Thomas Hawksley, *Replies to the Queries Issued by Her Majesty's Commissioners for Inquiring into the Present State of Large Towns* (London: William Clowes and Sons, 1845); John Billing, *The Necessity, Economy, and General Advantages of the Drainage of Towns by a Constant Supply of Water* (London: John Weale, 1847), 15; *Times,* 22 October 1844, 5; 25 March 1845, 4.

61. Councilor Dale speaking at the borough council meeting of 18 May 1852, quoted in *Bradford Observer,* 20 May 1852, 6; *Bradford Observer,* 22 July 1852, 8.

62. Councilor Dale speaking at the borough council meeting of 17 August 1852, quoted in *Bradford Observer,* 19 August 1852, 6. On that occasion, Dale's conclusions were supported by councilors Sharp and Shaw.

63. Water committee of the Glasgow Town Council, quoted in Burnet, *History of the Water Supply of Glasgow,* 58.

64. George Mason, engineer, quoted in John H. Smith, *Darlington, 1850* (Durham: Durham Historical Society, 1967), 6.

65. Proceedings of the Manchester Borough Council, 17 March 1847, quoted in Redford, *History of Local Government in Manchester,* 1:180.

66. The victims were Londoners. "Cholera in 1849," *Christian Remembrancer* 19 (1850): 165.

67. John Snow, "Cholera and the Water Supply in the South Districts of London," *British Medical Journal,* October 1857, 864.

68. Snow published *On the Mode of Communication of Cholera* at his own ex-

pense in 1849. Chadwick linked dysentery to poor drinking water; see Chadwick, *Report on the Sanitary Condition of the Labouring Population*, 149. Even after the Broad Street pump "experiment" of 1854, British medical professionals tended to blame miasmas, or airborne emanations of contagion, for most disease. See Christopher Hamlin, *Public Health and Social Justice in the Age of Chadwick, Britain, 1800–1854* (Cambridge: Cambridge University Press, 1998), 60–61.

69. *Bradford Observer*, 9 September 1852, 4.

70. The Lord Provost of the town council's moving acceptance of a chemist's report on the water supply in September 1853 is quoted in Burnet, *History of the Water Supply of Glasgow*, 72.

71. For its keen analysis of the ways in which humans have invested cholera with the power to force outcomes in society, see Christopher Hamlin, "'Cholera Forcing': The Myth of the Good Epidemic and the Coming of Good Water," *American Journal of Public Health* 99 (November 2009): 1946–54. For an extended consideration of cholera as just one factor in a complicated, multicausal story of public health development in this period, see Hamlin, *Cholera: The Biography* (Oxford: Oxford University Press, 2009), chap. 3 and elsewhere.

72. For his most accessible depiction of actor-network theory, see Bruno Latour's *Reassembling the Social* (Oxford: Oxford University Press, 2005). Vanessa Taylor and Frank Trentmann caution scholars to recognize as well the influences of consumer habits and political demands in determining water provision policies, especially regarding constant service and bath charges. New technologies and networks, they claim, created new opportunities for consumer activism, a picture that does not easily accord with certain depictions of governmentality. Their point is well taken. Vanessa Taylor and Frank Trentmann, "Liquid Politics: Water and the Politics of Everyday Life in the Modern City," *Past and Present* 211 (May 2011): 210–41.

73. *Ipswich Journal*, 11 December 1847, 3.

74. Mayor William Winterton, quoted in *Leicester Chronicle*, 3 April 1875, quoted in Elliot Malcolm, *Victorian Leicester* (London: Phillimore, 1979), 129.

75. Thomas Avery, *The Corporation of Birmingham and the Water Supply of the Town: A Statement Addressed to the Members of the Town Council* (Birmingham: The Journal, 1869), 15. Avery also reaffirmed the general belief that poor water supplies were at the root of cholera and that existing sources for commercial suppliers were unacceptable. Ibid., 8. Avery was later to become chair of the council's water committee.

76. W. T. Gairdner, *Public Health in Relation to Air and Water* (Edinburgh: Edmonston and Douglas, 1862), 150, 186–87. William Humber, writing a survey of the state of water supply in England, restated this general principle: "As to the quantity supplied, so with the quality—it is mostly a question of

dividend rather than what is really best for the consumers." Humber, *A Comprehensive Treatise on the Water Supply of Cities and Towns* (London: Crosby Lockwood, 1876), 86.

77. Gairdner, *Public Health,* 150.

78. Arthur Silverstone, *The Purchase of Gas and Water Works, with the Latest Statistics of Municipal Gas and Water Supply* (London: Crosby Lockwood, 1881), 69.

79. Stern, "Water Supply in Britain," 1002; Karen J. Bakker, *An Uncooperative Commodity: Privatizing Water in England and Wales* (Oxford: Oxford University Press, 2003), 52.

80. Stern, "Water Supply in Britain," 999, 1002; Hassan, "Growth and Impact of the British Water Industry," 534.

81. Sheard, "Nineteenth Century Public Health," 98.

82. Bakker, *Uncooperative Commodity,* 52–53.

83. John Hassan, *A History of Water in Modern England and Wales* (Manchester: Manchester University Press, 1998), 18–19, 134.

84. *Hansard Parliamentary Debates,* 3rd ser., vol. 314 (April 1887), cols. 150–68.

85. Sheffield's efforts to establish a municipal waterworks finally succeeded in 1887. Graham Taylor, *State Regulation and the Politics of Public Service: The Case of the Water Industry* (London: Mansell, 1999), 37.

86. Fraser, *Power and Authority,* 167.

87. Bakker, *Uncooperative Commodity,* 46–50.

88. Robert Millward, "Privatisation in Historical Perspective: The UK Water Industry," in *Money, Trade and Payments: Essays in Honour of D. J. Coppock,* ed. David Cobham, Richard Harrington, and George Zis (Manchester: Manchester University Press, 1989), 202.

89. Fraser, *Power and Authority,* 165.

90. *Bradford Observer,* 26 August 1852, 4.

91. Reverend Dr. Robert Buchanon, quoted in Irene Maver, *Glasgow* (Edinburgh: Edinburgh University Press, 2000), 89.

92. "Sanitary Reform," *Edinburgh Review* 91 (January 1850): 220, 213.

93. Avery, *Corporation of Birmingham and the Water Supply,* 14.

94. Henry Mayhew, *London Labour and the London Poor,* vol. 1 (London: George Woodfall, 1851), 101.

95. George Sims, *How the Poor Live* (London: Chatto and Windus, 1883), 39. Peter Thorsheim quotes Sims in his fine *Inventing Pollution: Coal, Smoke, and Culture in Britain since 1800* (Athens: Ohio University Press, 2006), 44. Thorsheim finds that many observers were concerned about the ways that air pollution could degrade the morals of city dwellers.

96. *Observer,* 28 October 1849, 3.

97. House of Commons, *Commission on the Employment of Children, Young Persons, and Women in Agriculture, 1867,* Second Report, part 1, 540; *Preliminary Report from Her Majesty's Commissioners on Agriculture,* C. 2778 (1881), 38.

98. *Observer,* 12 November 1849, 3.

99. Royal Commission for Inquiring into the State of Large Towns and Populous Districts, *First Report of the Commissioners for Inquiring into the State of Large Towns and Populous Districts* (1844), 81.

100. See John Broich, "Engineering the Empire: British Water Supply Systems and Colonial Societies, 1850–1900," *Journal of British Studies* 46 (April 2007): 346–65.

101. Huddersfield was one of many places where municipalization and expansion of works were to take place concomitantly. Woodhead, *History of the Huddersfield Water Supplies,* 57–61.

102. Silverstone, *Purchase of Gas and Water Works,* 88–125; Local Government Board, *Urban Water Supply: Return Showing the Means by Which Drinkable Water Is Supplied to Every Urban Sanitary District in England,* Wales, 3 July 1879 (London: HMSO, 1879), 2–143; Russell, "John Frederic La Trobe-Bateman," 199–200, 211.

103. *Medical Times and Gazette,* 3 October 1868, 405.

104. *History of the Water Supply of Newcastle,* 3, 6. This volume reprints a series of articles from the *Newcastle Chronicle* on the history of the town's water supply.

105. Dale Thomas, *On the Supply of Water to the Lancashire and Yorkshire Towns from the Lake Districts of Cumberland and Westmoreland* (London: Kent and Co., 1866), 3–4.

106. Alderman Murgatroyd, speaking in a town council meeting of 21 September 1852, quoted in *Bradford Observer,* 23 September 1852, 6.

107. Avery, *Corporation of Birmingham and the Water Supply,* 15; Lord Provost of the Glasgow Town Council, speaking in the council meeting of 15 September 1858, quoted in *Glasgow Herald,* 16 September 1858, 5.

108. Mayor C. H. Jones, speaking to the Huddersfield Town Council, quoted in *Huddersfield Chronicle,* 26 September 1868, 6. Jones was concerned that other towns might claim the new water source he wanted for Huddersfield.

109. Quoted in Burnet, *History of the Water Supply of Glasgow,* 154.

110. *Bradford Observer,* 16 March 1854, 4.

111. Gairdner, *Public Health,* 235.

112. "Supply of Water to the Metropolis," *Edinburgh Review* 91 (April 1850): 404.

113. Ibid., 405.

114. See, for example, Thomas Brazill, *Report to the Corporation of Dublin on the Proposed Supply of the City and Suburbs with Pure Water at High Pressure*

(Dublin: John Falconer, 1854), 5; John Frederic Bateman, *On the Present State of Our Knowledge on the Supply of Water to Towns* (London: Taylor and Francis, 1856), 63.

115. Bateman, *On the Present State of Our Knowledge on the Supply of Water,* 63.

116. Silverstone, *Purchase of Gas and Water Works,* 88–125. Liverpool and Birmingham were about to spend many millions more on their long-distance projects: the Vyrnwy and Elan valley schemes.

117. Silverstone, *Purchase of Gas and Water Works,* 98.

118. Redford, *History of Local Government in Manchester,* 2:186; John Frederic La Trobe-Bateman, *History and Description of the Manchester Waterworks* (Manchester: T. J. Day, 1884).

119. *Report of the Royal Commission on Water Supply* (London: HMSO, 1869), xv. Queen Victoria herself had opened the aqueduct in 1859 and had commemorated its advocates in a speech on the occasion. *Times,* 15 October 1859, 9. Glasgow had municipalized its water supply in 1852; see Maver, *Glasgow,* 91.

120. Silverstone, *Purchase of Gas and Water Works,* 72.

121. Leicester, Rochdale, and other towns built Hawksley gravitation schemes. See Roney, "Trial and Error," 80; *Leicester Water Undertaking,* 59; and Rebe Taylor, *Rochdale Retrospect* (Rochdale: Rochdale Borough Council, 1955), 127–28. Simpson promoted the system for Aberdeen and Liverpool. See James Simpson, *Report on the Most Efficient Means of Obtaining an Increased Supply of Water to the City of Aberdeen* (Aberdeen: George Cornwall, 1855), 11; and James Simpson and James Newlands, *Liverpool Water Supply, Report* (Liverpool: D. Marples, 1849), 10, 34, 48–50.

122. It is difficult to ascertain the exact number of Bateman's gravitation systems. Even his only biographer can offer only a range of the number of reservoirs he completed (seventy to eighty), and some of those were not for drinking water. Bateman also promoted some schemes that were adopted by town councils but never completed. I derive my figure of fifty from Russell, "John Frederic La Trobe-Bateman," 176, 182, 199–200, 211, 263; and J. F. Bateman, "On a Constant Water Supply for London: A Paper Read at a Meeting of the Health Department of the National Association for the Promotion of Social Science" (1867), Institution of Civil Engineers, London, 1. Geoffrey Binnie counts thirty completed. Geoffrey Binnie, "The Evolution of British Dams," in *Dams,* ed. Donald C. Jackson (Aldershot: Ashgate Variorum, 1998), 94.

123. Russell, "John Frederic La Trobe-Bateman," 113–14, 116–17.

124. Ibid., 133. Chadwick, of course, was not solely responsible for Bateman's conversion to professional water supply engineer, but Bateman's biographer attributes much influence to the activity of the Health of Towns Association and the Health of Towns Commission of 1844–45. Ibid., 105. Prior to 1844,

Bateman had promoted only one project devoted solely to drinking water; the project served Brighton and was completed in 1849. Ibid., 124.

125. Russell, "John Frederic La Trobe-Bateman," 176, 182, 211.

126. See, for examples, the description of the "greatest water scheme that has ever been devised," in Silverstone, *Purchase of Gas and Water Works,* 88. Bateman's plan for Bradford was to increase the daily supply from 576,000 to 10 million to 12 million gallons. *Bradford Observer,* 21 October 1852, 6.

127. Russell, "John Frederic La Trobe-Bateman," 198. See also the Lord Provost of Glasgow, speaking to the Glasgow Town Council meeting of 15 September 1853, quoted in *Glasgow Herald,* 16 September 1853, 5; and Alderman Farrar of the Bradford Town Council, speaking at a meeting of 19 October 1852, quoted in *Bradford Observer,* 21 October 1852, 6.

128. Maver, *Glasgow,* 91.

129. Queen Victoria's speech, quoted in *Times,* 15 October 1859, 9.

130. Edward Filliter, *Report on the Best Mode of Obtaining an Additional and Purer Supply of Water for the Borough of Leeds* (Leeds: Charles Goodall, 1866), 7.

131. Filliter had recommended accumulating water from the upland river Burn north and west of Leeds. In 1880, Bradford welcomed an act allowing the city to gather water from the valley of the river Nidd and transport it more than thirty miles. Filliter, *Report on the Best Mode of Obtaining an Additional and Purer Supply of Water for the Borough of Leeds,* 30–31.

132. Bateman, *On the Present State of Our Knowledge on the Supply of Water,* 72, 75. See, for example, Thomas, *On the Supply of Water to the Lancashire and Yorkshire Towns,* 5.

133. Brazill, *Report to the Corporation of Dublin,* 4. Brazill proposed damming a deep valley near the source of the river Liffey, creating a reservoir fifty feet deep and one mile long; the water would travel twenty-two miles in an aqueduct to the city. Ibid., 8–9, 12.

134. Hassan, "Growth and Impact of the British Water Industry," 545. A small reservoir, probably on the scale of a millpond, had been built in Greenhill in 1796; it failed within fifteen years. In 1882, the Manchester and Salford Waterworks built a small reservoir (at least partially used for drinking water) with a dam thirty-three feet high. Russell, "John Frederic La Trobe-Bateman," 59.

135. G. M. Binnie, *Early Victorian Water Engineers* (London: Telford, 1981), 9–13, 130–31.

136. The Chelsea Company allotted 19 percent of the houses it served a constant supply; the East London Company, 85 percent (though it is uncertain how many houses had the necessary fittings to receive it); the Grand Junction, 78 percent; Kent, 53 percent; Lambeth, 51 percent; New River, 37 percent; Southwark and Vauxhall, 43 percent; and West Middlesex, 34 percent. H. C.

Richards and W. H. C. Payne, *London Water Supply*, 2nd ed. (London: P. S. King, 1899), 94.

137. Simpson and Newlands, *Liverpool Water Supply*, 7.

138. Bateman, "On a Constant Water Supply for London," 1.

139. Thomas, *On the Supply of Water to the Lancashire and Yorkshire Towns*, 9.

140. Thomas's scheme—never built—would have supplied Leeds, Bolton, and Liverpool, with its main trunk extending 170 miles. Thomas, *On the Supply of Water to the Lancashire and Yorkshire Towns*, 18.

141. Bateman, *History and Description of the Manchester Waterworks*, 206.

142. Health of Towns Association, *Appendix II, Second Report of the Commissioners for Inquiring into the State of Large Towns and Populous Districts* (London: Clowes and Sons, 1845), 39.

143. Health of Towns Association, *Report of the Sub-Committee on Answers to Questions Addressed to the Principal Towns of England and Wales* (London: Brewster and West, 1848), 16.

144. Cuthbert Johnson, *The Acts for Promoting the Public Health, 1848–1851* (London: Charles Knight, 1852), 240.

145. *Edinburgh Annual Register, 1824* (Edinburgh: James Ballantyne and Co., 1825), 236.

146. *Mechanic's Magazine* 30 (30 March 1839): 452.

147. *Mechanic's Magazine* 38 (7 January 1843): 168.

148. *London Quarterly Review* 89 (March 1851): 253.

149. Metropolitan Working Classes' Association for Improving the Public Health, *Water Supply, Especially for the Working Classes* (London: John Churchill, 1847), 7. Despite its name, this association was led by London patricians, not working-class Londoners. Its board comprised ecclesiasts, members of the House of Lords, Royal Society fellows, and so on; see Samaritan Fund Committee of St. George's and St. James's Dispensary, *Ventilation Illustrated* (London: John Churchill, 1848), 37.

150. House of Commons, *Report of the Select Committee on Fires in the Metropolis, 1862*, 147.

151. Ibid., 90, 147.

152. Ibid., 90.

153. Ibid., 111–12.

154. Mayhew, *London Labour and the London Poor*, vol. 2 (London: Griffin, Bohn, and Co., 1861), 78.

155. Health of Towns Association, *Report of the Sub-Committee on Answers to Questions*, 45.

156. Health of Towns Association, *Appendix II, Second Report of the Commissioners*, 273.

157. *Mechanic's Magazine* 38 (7 January 1843): 166, 173.

158. Hassan, "Growth and Impact of the British Water Industry," 540.

159. This amount was Exeter's budget for its local board of health in 1870–71. Robert Newton, *Victorian Exeter, 1837–1914* (Leicester: Leicester University Press, 1968), 286.

160. Burnet, *History of the Water Supply of Glasgow,* 133; Silverstone, *Purchase of Gas and Water Works,* 89.

161. Silverstone, *Purchase of Gas and Water Works,* 89, 100.

162. Gemmel, speaking in a town council meeting of 27 September 1853, quoted in *Glasgow Herald,* 28 September 1853, 5.

163. Townsperson Mr. Senior, quoted in Russell, "John Frederic La Trobe-Bateman," 207.

164. Fraser, *Power and Authority,* 46.

165. Avery, *Corporation of Birmingham and the Water Supply,* 23. As of 1881, Birmingham charged for water on the same scale as the erstwhile water company. Silverstone, *Purchase of Gas and Water Works,* 89.

166. Roney, "Trial and Error," 109.

167. Economic historian John Hassan, too, finds this to be the case. Hassan, *History of Water in Modern England and Wales,* 19.

168. See, for example, the extended pamphlet by the leader of the local Conservative Association in Liverpool: Samuel Holme, *Want of Water: A Letter to Harmood Banner, in Reply to His Pamphlet Entitled "Water"* (Liverpool: Lace and Addison, 1845). See also Roney, "Trial and Error," 109; *Bradford Observer,* 16 March 1854, 4.

169. See Fraser, *Power and Authority.*

170. Briggs, *Victorian Cities,* 193–94.

171. Sheard, "Nineteenth Century Public Health," 100.

172. Walter Stern puts the number of members of Parliament (MPs) who had financial interests in water companies in the early 1850s at between seventy and ninety. Stern, "Water Supply in Britain," 999.

173. *Leeds Mercury,* 8 June 1866, 3.

174. *Liverpool Mercury,* 1 January 1875, 6. While there is no evidence that the closing of wells in favor of municipal water supplies led to vehement protests in Britain on the grounds of tradition, this example does provide an interesting comparison to colonial cases of protests against well closings. See Broich, "Engineering the Empire," 360.

175. Great Britain, Royal Commission on Labour, *The Agricultural Labourer* (London: HMSO, 1893), 118.

176. I. Davidson, "George Deacon (1843–1909) and the Vyrnwy Works," in *Dams,* ed. Jackson, 171.

177. For details of the Vyrnwy project, see ibid., 171–86.

178. Davidson, "George Deacon (1843–1909) and the Vyrnwy Works," 172.

179. James C. Scott, *Seeing Like a State: How Certain Schemes to Improve the Human Condition Have Failed* (New Haven: Yale University Press, 1998).

Chapter 2. Great Expectations

1. *Supply of Water to the Metropolis: Plenty, Purity, Pressure, Price* (London: Trelawney Saunders, 1849), 5.

2. "Supply of Water to the Metropolis," *Edinburgh Review* 91 (April 1850): 385.

3. Henry Morley, "A Home Question," *Household Words* 10 (11 November 1854): 295.

4. Transcript of proceedings of a public meeting held at Hanover Square Rooms, 22 October 1849, 1, Sanitary Reform of London Collection, Stanford University Library Special Collections, Stanford, CA.

5. Asa Briggs, *Victorian Cities* (New York: Harper and Row, 1965), 85–86, 312.

6. Roy Porter, *London: A Social History* (Cambridge, MA: Harvard University Press), 205.

7. These figures are derived from Charles Creighton, *History of Epidemics in Britain,* 2 vols. (London: Cassell, 1894; repr., London: Cass, 1965), 2:820–35, 840–48, 851–45, 856–59.

8. Anne Hardy, "Water and the Search for Public Health in London in the Eighteenth and Nineteenth Centuries," *Medical History* 28, no. 3 (1984): 251–52.

9. Beer consumption per capita remained high until 1900 but trailed downward from that point. See F. M. L. Thompson, *The Rise of Respectable Society: A Social History of Victorian Britain, 1830–1900* (Cambridge, MA: Harvard University Press, 1988), 308–13.

10. By the mid-nineteenth century, the New River Company had purchased water companies that drew from the Thames but still banked on its reputation for supplying superior water from the hinterlands. See, for example, John Hogg, *London as it is; being a series of observations on the health, habits, and amusements of the people* (London: John Macrone, 1837), 257.

11. For 1700, John Graham-Leigh lists suppliers as the London Bridge, New River, Chelsea, and York Buildings Companies north of the river in the built-up area proper and the Hampstead and Shadwell Companies in the suburbs. John Graham-Leigh, *London's Water Wars: The Competition for London's Water Supply in the Nineteenth Century* (London: Francis Boutle, 2000), 12–13. For the 1800 data, see Nicola Tynan, "Private Water Supply in Nineteenth Century London: Re-assessing the Externalities," National Bureau of Economic Research Summer Institute–Development of the American Economy, Cambridge, MA, July 2000, 39–40.

12. Stephen Inwood, *A History of London* (New York: Carroll & Graf, 1998), 5.

13. Graham-Leigh, *London's Water Wars,* 24–28, 95–96; Hardy, "Water and the Search for Public Health," 252.

14. Hardy, "Water and the Search for Public Health," 253.

15. Remarkable detail on these matters is in Graham-Leigh, *London's Water Wars*, 25. The West Middlesex Company determined to use iron pipes from its inception around 1808; other companies only slowly followed its lead. Ibid., 27–28.

16. Hardy, "Water and the Search for Public Health," 252.

17. Graham-Leigh, *London's Water Wars*, 19; Hardy, "Water and the Search for Public Health," 256.

18. Asok Kumar Mukhopadhyay, *Politics of Water Supply: The Case of Victorian London* (Calcutta: World Press, 1981), 4.

19. The East London, New River, London Bridge, Grand Junction, West Middlesex, York Buildings, and Chelsea Companies cooperated; the Kent and Vauxhall Companies lacked competition in their areas. See Graham-Leigh, *London's Water Wars*, 52–61.

20. Stephen Halliday, *The Great Stink of London: Sir Joseph Bazalgette and the Cleansing of the Victorian Capital* (Stroud: Sutton, 1999), 45, 46.

21. On the adoption of water closets, see Dale Porter, *The Thames Embankment: Environment, Technology, and Society in Victorian London* (Akron: University of Akron Press, 1998), 55–56; Halliday, *Great Stink*, 42–48.

22. David Sunderland, in the course of making an argument that London's water companies in the period did not provide poor service relative to the challenges they faced, alleges a bit of a conspiracy between Wright and a former director of the Grand Junction Water Company. Sunderland argues that the retired director had persuaded Wright to author the muckraking pamphlet as an act of revenge against the company. But given the testimony of horrible water supplies aired by many area residents in a public meeting of 9 April 1827, a conspiracy hardly seemed necessary. David Sunderland, "'Disgusting to the Imagination and Destructive of Health'? The Metropolitan Supply of Water, 1820–52," *Urban History* 30, no. 3 (December 2003): 371. On the public meeting, see *Times*, 10 April 1827, 3.

23. John Wright, *The Dolphin, or Grand Junction Nuisance, Proving the Several Thousand Families in Westminster and its Suburbs are Supplied with Water in a State Offensive to the Sight, Disgusting to the Imagination and Destructive to Health* (London: T. Butcher, 1828), 61.

24. Wright included doctors' testimony that the water was unfit for drinking and other domestic purposes. Wright, *Dolphin*, part III, sec. 9.

25. *Times*, 9 April 1827, 2.

26. *Times*, 10 April 1827, 3.

27. Royal Commission on Metropolitan Water Supply, 1828, Minutes of Evidence (London: HMSO, 1828), 125.

28. *Report of the Royal Commission on Metropolitan Water Supply, 1828* (London: HMSO, 1828) 12.

29. Ibid. Chadwick referred to the findings of the 1828 Royal Commission on Metropolitan Water Supply in his famed *Report on the Sanitary Condition of the Labouring Population of Great Britain* (London: Clowes and Sons, 1842).

30. The shocking points of *The Dolphin* were repeated in the *Times,* 19 March 1827, 2; it also reprinted the indictments of expert witnesses before the commission. *Times,* 15 March 1828, 3.

31. Report of the Select Committee on the Supply of Water to the Metropolis, 1821, Minutes of Evidence (London: HMSO, 1821), 7.

32. George Cruikshank, *Salus Populi Suprema Lex,* 1832. This broadside is included in M. Dorothy George, *Catalogue of Political and Personal Satires Preserved in the Department of Prints and Drawings in the British Museum,* vol. 11 (London: Trustees of the British Museum, 1954), which nicely transcribes all the things the figures are saying in the broadside.

33. Hogg, *London as it is,* 255.

34. "Supply of Water to the Metropolis," *Edinburgh Review* 91 (April 1850): 386.

35. MP Charles Lushington, transcript of public meeting, Hanover Square Rooms, 22 October 1849, 12–13, 15.

36. Arthur Hill Hassall, *A Microscopic Examination of the Water Supplied to the Inhabitants of London and the Suburban Districts* (London: Samuel Highley, 1850), 59.

37. Rev. John Garwood, quoted in W. Archdall O'Dougherty, *Water for Domestic Use: Evils Attending the Use of Impure Water, to Health, to Purse, and to Morals* (n.p., 1862), 21. This source quotes many reformers on what they viewed as the close connection between poor-quality water and alcohol abuse among the poor. Also, Edwin Chadwick linked insufficient water to the dirty state of the masses and their quarters, and the dirty state of one's quarters to the disgrace of the working-class individual, and that disgrace to the predilection to drink. He reported in 1842 that the "undrained" abode "has an effect on the moral habits by acting as a strong and often irresistible provocative to the use of fermented liquors." Chadwick, *Report on the Sanitary Condition of the Labouring Population,* 130.

38. *Hansard Parliamentary Debates,* 3rd ser., vol. 249 (August 1879), col. 918.

39. To cite just a handful of examples, a writer to the *Northern Echo* argued that the temperance movement should work first on securing pure, abundant water supplies so as to "raise an increasing army of water drinkers." *Northern Echo* (Darlington), 8 November 1893, 2. MP John Stapleton argued that the East River Company, by not pumping water on Sundays, was a particular threat to morals because on the sabbath, Londoners would mix alcohol with their saved cistern water to make it more palatable, thus threatening their ability "to attend church decently." *Hansard Parliamentary Debates,* 3rd ser., vol. 200 (April 1870), col. 1368. The Church of England Temperance Society

encouraged the reform of London's water supply in the name of encouraging temperance and good morals, as well. *Church of England Temperance Chronicle* 8 (27 November 1880): 763.

40. "Supply of Water to the Metropolis," *Edinburgh Review*, 384.

41. John Morley, "Piping Days," *Household Words* 10 (14 October 1854): 197, 198.

42. Hassall, *Microscopic Examination of the Water Supplied*, 59.

43. *Supply of Water to the Metropolis: Plenty, Purity*, 12.

44. Hogg, *London as it is*, 261.

45. "Supply of Water to the Metropolis," *Edinburgh Review*, 405.

46. *Times*, 21 April 1851, 4.

47. Hassall, *Microscopic Examination of the Water Supplied*, 59.

48. Transcript of public meeting, Hanover Square Rooms, 22 October 1849, 1.

49. *Second Report of the Commissioners for Inquiring into the State of Large Towns and Populous Districts* (London: HMSO, 1845), 50.

50. *Report by the General Board of Health on the Supply of Water to the Metropolis* (London: HMSO 1850), 1. Chadwick was one of the four authors of the report.

51. *Report by the General Board of Health on the Supply of Water*, 275.

52. See, for another example, *Supply of Water to the Metropolis: Plenty, Purity*, 4.

53. "By the new theory[,] rivers were altogether repudiated as sources of supply," reported the *Times*. Many even rejected the tributaries of the Thames. *Times*, 21 April 1851, 4.

54. "Supply of Water to the Metropolis," *Edinburgh Review*, 393, 404; *Supply of Water to the Metropolis: Plenty, Purity*, 12.

55. "Supply of Water to the Metropolis," *Edinburgh Review*, 404; *Report by the General Board of Health on the Supply of Water*, 100.

56. See, for example, Hassall, *Microscopic Examination of the Water Supplied*, 34.

57. *Correspondence between John Stuart Mill, Esq., and the Metropolitan Sanitary Association on the Proper Agency for Regulating the Water-Supply for the Metropolis* (London: J. Gadsby, 1851), 21. See also ibid., 22.

58. On London government in the first half of the century, see Inwood, *History of London*, 358–61. In 1850, most of the population of London was concentrated in approximately twenty significant parishes.

59. The designated location west of the city was above Teddington weir, near Kingston-upon-Thames, where the river's tides were halted by the dam.

60. See, for example, a letter to the *Times* calling for authorities "to prevent the sewage of Uxbridge from passing into the Thames close to the very spot whence the water for London is taken." *Times*, 18 June 1872, 12.

61. See, for example, William A. Robson, *The Government and Misgovernment of London*, 2nd ed. (London: George Allen and Unwin, 1948), 105; or David Owen, *The Government of Victorian London, 1855–1889* (Cambridge, MA: Harvard University Press, 1982), 134.

62. With his insistence on universal drainage, Chadwick knew that he was promoting the pollution of the river in order to reduce the pollution in the poor alleys. He wrote in his report that "the chief objection to the extension of this system is the pollution of the water of the river in which the sewers are discharged. Admitting the expediency of avoiding the pollution, it is nevertheless proved to be an evil of almost inappreciable magnitude in comparison with the ill health occasioned by the constant retention of pollution of several hundred thousand accumulating in the most densely peopled districts." Chadwick, *Report on the Sanitary Condition of the Labouring Population*, 120.

63. Chadwick stated that the harmful effects of sewage were halted by "putting the solid manure into water to arrest the gases." Edwin Chadwick, "Discussion on 'The Utilisation of the Sewage of Towns,'" *Journal of the Society of Arts* 212 (12 December 1856): 49. On the prevalence of miasma theory, see Porter, *Thames Embankment*, 55. See also M. W. Flinn's modern introduction to the Edinburgh University Press reprint edition of Chadwick, *Report on the Sanitary Condition of the Labouring Population*, 62–64.

64. Hall quoted in *Times*, 14 August 1855, 6.

65. To keep the proposed governing body an efficient size, smaller vestries would combine to send shared representatives. Since many vestries were not popularly elected by parish residents, the new board would also not be representative.

66. See the account of the creation of the MBW in Owen, *Government of Victorian London*, 31–38.

67. See Owen, *Government of Victorian London*, 43–44. Robert Stephenson, the founding father of railway engineering, William Cubitt, the eminent rail and waterway engineer, and the legendary I. K. Brunel offered testimonials on Bazalgette's behalf. See Halliday, *Great Stink*, 66.

68. The terms of the Metropolis Management Act of 1855 stipulated that the government's Office of Works must review, and Parliament must approve, any projects costing more than £100,000. See Porter, *Thames Embankment*, 70.

69. On the "Great Stink," see Porter, *Thames Embankment*, 71; or Halliday, *Great Stink*, ix–xi. On the struggles between the MBW and the government over the final design of the scheme, see Owen, *Government of Victorian London*, 50–52; and Porter, *Thames Embankment*, 70.

70. Porter, *Thames Embankment*, 157. The MBW was technically empowered to raise rates in the vestries but in 1858 doubted its ability to compel the vestries to collect and deliver them. Ibid., 151.

71. The duties on these imports had been tied up for other purposes for the previous thirty years. Porter, *Thames Embankment*, 155.

72. Porter, *Thames Embankment*, 155.

73. Select Committee on the Thames River (Prevention of Floods) Bill, Minutes of Evidence, 1877 (London: HMSO, 1877), 5.

74. The extant Abbey Mills lift station is extremely ornate, executed in a style sometimes called Venetian gothic.

75. The MBW imported stone from Cornwall, Ireland, Brittany, and other regional sources; see the discussion quoted in the *Proceedings of the Institution of Civil Engineers* 53 (9 April 1878): 2.

76. Odor remained closely related to contagion in the minds of many. William Odling, the health officer for Lambeth, was so convinced that the removal of the mud banks would clear the most dangerous aspects of Thames pollution that he believed the embankments' intercepting sewers were a waste of money. William Odling, *Report of the Effects of Sewage Contamination upon the River Thames* (Lambeth: G. Hill, 1858), 15.

77. W. Fothergill Cooke, "On the Utilisation of the Sewage of Towns," *Journal of the Society of Arts* 212 (12 December 1856): 55. Naturally, London's engineers, too, saw embankments on the Thames as an obvious mark of the progress of civilization. The *Proceedings of the Institution of Civil Engineers* recorded that "with the development of civilization [the Thames's] waters were restrained by embankments." *Proceedings of the Institution of Civil Engineers* 53 (9 April 1878): 2.

78. Guildford B. Richardson (MBW member since 1862) statement, Select Committee on the Thames River (Prevention of Floods) Bill, Minutes of Evidence, 1877, 147.

79. *Handbook to London as It Is* (London: John Murray, 1876), 42.

80. Ibid., 26.

81. *Times,* 9 December 1874, 7.

82. Edward J. Watherston, *The Water Supply of the Metropolis: Addresses to the Delegates from the Vestries and District Boards of the Metropolis* (London: Dryden Press, 1879), 7, 9.

83. William Rendle, *What Are the Advantages or Disadvantages of Water Supply Being Lodged in the Hands of Local Authorities?* (London: n.p., n.d.), 4.

84. Ibid., 16.

85. "The Water Supply of London," *Quarterly Review* 127 (1869): 450.

86. Ibid., 451; *Metropolitan Board of Works Report on the Provisions of the Metropolis Water Bill, 1871,* 2; *Report of the Royal Commission on Water Supply, 1869* (London: HMSO, 1869), cvi.

87. J. Bailey Denton, "On the Water Supply of the Metropolis, in Relation to the Conservancy of the Thames and Its Tributaries," *Journal of the Society*

of Arts 15 (June 1867): 467. Denton was writing to counter a perception in the public that the Thames would be exhausted. The use of reservoirs, he argued, would prevent that.

88. John Frederic Bateman, "On Constant Supply for London," paper read at the Meeting of the Health Department of the National Association for the Promotion of Social Science, January 1867, 1.

89. The new source could be supplemented by a regional supply drawn from wells, if necessary. *Report of the Royal Commission on Water Supply, 1869*, viii.

90. John Frederic Bateman, *On the Supply of Water to London from the Sources of the River Severn* (London: Vacher and Sons, 1865), 9.

91. *Report of the Royal Commission on Water Supply, 1869*, ix–x.

92. Bateman, *On the Supply of Water to London*, 13–14.

93. Bateman's mind had been distilling the idea since 1860, when he had finished his celebrated Glasgow works, and he had been actively making preliminary investigations and drawings since 1862. He began surveying the gathering grounds in 1865, as if he knew that London must soon commit to an importation scheme, and spent £3,000 of his own money to do so. Peter Ellerton Russell, "John Frederic La Trobe-Bateman, F.R.S., Water Engineer 1810–1889" (MS thesis, University of Manchester, 1980), 232.

94. Russell, "John Frederic La Trobe-Bateman," 232–33.

95. "We are of the opinion," the commissioners reported, "that there is no evidence to lead us to believe that the water now supplied by the companies is not generally good and wholesome." *Report of the Royal Commission on Water Supply, 1869*, cxxvi–cxxvii.

96. This story fits into a wider, compelling history of the development of water analysis expertise, as told in Christopher Hamlin, *A Science of Impurity: Water Analysis in Nineteenth Century Britain* (Berkeley and Los Angeles: University of California Press, 1990).

97. The companies were "prepared . . . to supply [180 million gallons per day]. Which is very little short of the quantity we have estimated as the highest demand that need be reasonably looked forward to for the metropolis." *Report of the Royal Commission on Water Supply, 1869*, cix. The commissioners' satisfactory conclusion was based on an eventual peak population of 5 million, while London's population, in fact, exceeded 5 million by 1890 and peaked somewhere near 8.7 million. Ibid., cvii. For London's population, see Porter, *London*, 306.

98. *Report of the Royal Commission on Water Supply, 1869*, cxxvi.

99. Quoted in *Journal of Gas Lighting, Water Supply, & Sanitary Improvement* 18 (6 July 1869): 538; 18 (17 August 1869): 658.

100. *Journal of Gas Lighting, Water Supply, & Sanitary Improvement* 18 (6 July 1869): 537–38.

101. Ibid., cxxvii.

102. Board member Hows, quoted in *Times,* 18 December 1869, 8.

103. Board member Dresser Rogers, quoted in *Times,* 18 December 1869, 8.

104. It bears repeating that Walter Stern puts the number of MPs invested in water companies in the early 1850s at between seventy and ninety. Walter M. Stern, "Water Supply in Britain: The Development of a Public Service," *Royal Sanitary Institute Journal* 74 (October 1954): 999. On the successful undermining of the 1871 act, see Owen, *Government of Victorian London,* 135–36. The act also created the position of the official metropolitan water examiner, who reported to the Local Government Board once monthly. See Hamlin, *Science of Impurity,* esp. chap. 7.

105. Vestries south of the river, served by the Southwark and Vauxhall Company and in the east, served by the East London Company, were particularly vocal in this period. Metropolitan Board of Works Minutes of Proceedings, January–June 1877 (London: Reed and Pardon, 1877), 302, 354, 391, 466–7, 524, 540, 594.

106. In general, see J. W. Bazalgette, F. J. Bramwell, and E. Easton, *Report upon an efficient supply of water to the Metropolis,* 10 August 1877. The engineers F. J. Bramwell, a fellow of the Royal Society and later knighted, and Edward Easton assisted Bazalgette. See also *Metropolitan Board of Works Annual Report, 1877* (London: Reed and Pardon, 1877), 76–77.

107. Consulting engineer F. J. Bramwell describing the chalk water scheme in a speech before the British Association for the Advancement of Science, transcribed in *Times,* 22 August 1877, 11.

108. Ibid.

109. That is, 16 million out of 120 million gallons.

110. Bateman, *On the Supply of Water to London,* 18.

111. *Metropolitan Board of Works Annual Report, 1878,* 64.

112. *Times,* 11 August 1877, 10; 22 August 1877, 9.

113. *Metropolitan Board of Works Annual Report, 1878,* 64–65. Unsigned flyers appeared bearing cartoons mocking the awkwardness of having two water taps in each house. See item 72, Sanitary Reform of London Collection, Rare Books, Stanford University Library Special Collections, Stanford, CA.

114. *Hansard Parliamentary Debates,* 3rd ser., vol. 238 (1878), col. 1227.

115. Ibid., col. 1229.

116. Alderman Cotton, in ibid., col. 1231.

117. Henry Fawcett, in *Hansard Parliamentary Debates,* 3rd ser., vol. 238 (1878), col. 1232.

118. *Hansard Parliamentary Debates,* 3rd ser., vol. 238 (1878), col. 1234.

119. *Times,* 27 August 1877, 10.

120. *Hansard Parliamentary Debates,* 3rd ser., vol. 201 (1870), col. 1496.

121. James Beal, *Municipal Corporation for the Metropolis* (London: G. Phipps, 1862), 1.

122. Ibid., 2–3.

123. Ibid., 10. For Beal's extended argument about water supply, including a scolding for lagging behind foreign capitals, see ibid., 15–16.

124. Joseph F. B. Firth, *Municipal London* (London: Longmans, Green, & Co., 1876), 698.

125. Ibid., 391.

126. On the charges hanging over the MBW, see John Davis, *Reforming London: The London Government Problem, 1855–1900* (Oxford: Oxford University Press, 1988), 110. See also Owen, *Government of Victorian London*, 183–90.

127. Owen, *Government of Victorian London*, 42.

128. London governmental reform was on the minds of many in Prime Minister William Gladstone's government, as it was intermittently throughout the century, but the specific issue of London reform was dropped in favor of county-level reform under the new Conservative government of 1886. See Davis, *Reforming London*, 101–2.

129. In terms of having its own sheriff and bench of magistrates, London had been a county for several centuries by this time.

130. "Lodgers," who were single renters of a single room or a few rooms, were not qualified to vote, but a head of a household, even renting the least expensive cellar room, was qualified; see Albert Shaw, "How London Is Governed," *Century Illustrated Magazine*, November 1890, 141. On the high number of female voters, see Susan D. Pennybacker, *A Vision for London, 1889–1914: Labour, Everyday Life and the LCC Experiment* (New York: Routledge, 1995), 11.

131. For Beal's perception that it would be a straightforward matter for the new county council to municipalize, see an interview of Beal by unknown interviewer, 1 February 1889, F/BL/12/11, James Beal Papers, London Metropolitan Archives.

132. See also Beal's notes for a campaign speech, 1889, F/BL/05, Beal Papers. He coupled gas municipalization efforts with those for water.

133. Minutes of LCC meeting, 19 March 1889, London County Council Minutes of Proceedings, January–December 1889. See also *First Annual Report of the Proceedings of the Council*, 1890 (London: LCC, 1890), 76.

134. See the *Report of the Royal Commission on the Pollution of Rivers, 1874* (London: HMSO 1874); *Report of the Select Committee on Metropolitan Fire Brigade, 1876* (London: HMSO, 1876). On the 1880 committee, commonly referred to as Lord Cross's Committee, and its motivations, see Owen, *Government of Victorian London*, 139. The 1880 committee's recommendation to purchase the water companies came to nothing when calculations showed that the purchase price would be astronomical.

135. *Times,* 7 August 1879, 9. "The London Water Supply Question . . . is now

approaching a crisis, and it is clear that its settlement cannot be much longer postponed," predicted an anonymous author in 1881. The writer continued, noting that "the unhappy consumers of water under the . . . rule of the London Water Companies . . . have been . . . long accustomed to the miseries, discomforts, and uncleanliness which bad water always entails." "Twenty Years' Practical Experience," *The Truth about the London Water Supply Question* (London: E. W. Allen, 1881), 3, 17.

136. *Times*, 21 July 1890, 4.

137. Shaw, "How London Is Governed," 144–45. Albert Shaw had written an article on the government of Glasgow; see Albert Shaw, "Glasgow: A Municipal Study," *Century Magazine* 39 (March 1890): 721–36.

138. *Financial News*, 20 October 1890, 2.

139. J. F. B. Firth was the first acting chair of the Special Committee on Water Supply until Beal was elected. Beal was joined by Henry Clark, W. B. Doubleday, Sir Thomas Farrer, Frederic Harrison, Alfred Hoare, Alfred Jordan Hollington, W. J. P. Hunter, William Johnson, T. W. Maule, W. Phillips, and Captain J. Sinclair.

140. Sir T. H. Farrer, "A Model City," *New Review* 4 (1891): 213.

141. Frederic Harrison lecture, "The Functions of the State," 1893, reprinted in Harrison's *On Society* (London: Macmillan, 1918), 72. The *Bradford Observer* had argued in 1852 that water "ought to be as free as air." *Bradford Observer*, 26 August 1852, 4.

142. Minutes of the London County Council, 15 May 1889, 288–90, LCC/MIN/6257; 15 June 1889, 1–2, LCC/MIN/12,267, London Metropolitan Archives. The number of companies was reduced by midcentury due to consolidation.

143. Beal letter, 6 October 1890, LCC/MIN/12,267.

144. Farrer himself used the term "water reform." Farrer, "A Model City," 228.

145. Alexander Binnie, letters to the LCC Water Committee, October 1890 and 15 December 1890, LCC/MIN/12,267.

146. The first London County Council was not as rigidly divided along partisan lines as were subsequent sessions. Mukhopadhyay, *Politics of Water Supply*, 81. See also ibid., 76.

147. See, for example, letters from the St. James vestry (5 June 1890) and St. Olave District Council (27 April 1891), LCC/MIN/12,267; and from the Lambeth (13 December 1891), St. George (7 December 1891), and St. Luke Middlesex (7 January 1892) vestries, LCC/MIN/12,268, London Metropolitan Archives.

148. An example of the continued support is in *Times*, 11 September 1890, 7.

149. See refutations of some misconceptions on this point in London government historiography in Davis, *Reforming London*, 105–6.

150. Porter, *London,* 156–58. On Parliament's fear of a powerful body that might resist its programs, see also Gloria C. Clifton, *Professionalism, Patronage and Public Service in Victorian London* (London: Athlone Press, 1992), 22 and elsewhere.

151. Michael Bentley, *Lord Salisbury's World: Conservative Environments in Late-Victorian Britain* (Cambridge: Cambridge University Press, 2001), 37, 90.

152. Beal, *Municipal Corporation for the Metropolis,* 10.

153. Davis, *Reforming London,* 103.

154. Briggs, *Victorian Cities,* 331.

155. Even then, the LCC had only the power to form an understanding with the companies; Parliament, under the terms of the General Powers Act of 1890, which had made possible the formation of the London County Council, would have to approve the terms of purchase.

156. Mukhopadhyay, *Politics of Water Supply,* 76.

157. W. H. Dickinson, "The Water Supply of London," *Contemporary Review* 71 (February 1897): 234; "The London Water Companies, Their Rights and Dangers," *Investors' Review* 4 (1895): 206.

158. [Matthew White Ridley], *Select Committee on Metropolitan Water Supply, 1891* (London: HMSO, 1891); Mukhopadhyay, *Politics of Water Supply,* 79.

159. Sir Thomas Farrer, speaking in LCC debate, 13 October 1891, reprinted in whole in William Saunders, *History of the First London County Council* (London: National Press Agency, 1891), 530–31.

160. Ibid., 529.

161. The Royal Commission on Metropolitan Water Supply was chaired by Lord Balfour of Burleigh and included Sir G. B. Bruce, Sir A. Geikie, James Dewar, G. H. Hill, James Mansergh, and William Ogle. It convened in January 1892 and reported in September 1893.

162. Among those who argue that state intervention was increasing in this period is Ellen Frankel Paul, "Laissez Faire in Nineteenth-Century Britain: Fact or Myth?" *Literature of Liberty* 3 (winter 1980): 31–32, 34.

Chapter 3. "Communism in Water"

1. Christopher Hamlin, "The 'Necessaries of Life' in British Political Medicine, 1750–1850," *Journal of Consumer Policy* 29 (December 2006): 373–97.

2. "Governmental" here means the Foucauldian sense of a set of society-influencing practices that extend well beyond governmental policy.

3. Showing that this was not an inevitable path was the goal of Christopher Hamlin in *Public Health and Social Justice in the Age of Chadwick: Britain, 1800–1854* (Cambridge: Cambridge University Press, 1998). Mid-nineteenth-century Britain saw tremendous population growth and urbanization, with epidemic following close behind. And the individual who did the most to shape government and technical responses was Edwin Chadwick, utilitarian and author of

the famous *Report on the Sanitary Condition of the Labouring Population of Great Britain* (1842). He was instrumental, Hamlin explains, in defining government's role as limited in terms of both technical scope and social ambition. In other words, Chadwick and his allies helped make waterworks, sewer systems, and their administration merely sustain the status quo of industrial Britain.

4. Hamlin, *Public Health and Social Justice*, 214, 340, and elsewhere.

5. Municipalization advocates pointed out that European and American cities were constructing impressive waterworks of their own. The *Edinburgh Review* indicated that "it will be disgraceful to the nation if . . . we are surpassed in the arrangements for securing health and common decency, not only by the young republics of the New World, but even the ancient empires of the Old. We boast of our wealth, our freedom, our science, . . . and our love of cleanliness; we glory in our civilization, but our glory becomes our shame, if still we are last in the race of humanity. The City of New York has expended 2,500,000 on the Croton Water Works . . . an exertion of which our Transatlantic brethren may well be proud." *Edinburgh Review* 91 (1850): 405. Indeed, advocates of new waterworks frequently pointed to New York's Croton scheme, completed in 1842, as the model project: large, far sighted, beyond merely sufficient. See, for example, Thomas Brazill, *Report to the Corporation of Dublin on the Proposed Supply of the City and Suburbs with Pure Water at High Pressure* (Dublin: John Falconer, 1854), 5; John Frederic Bateman, *On the Present State of Our Knowledge on the Supply of Water to Towns* (London: Taylor and Francis, 1856), 63; *Manchester Guardian*, 4 December 1844, 6.

6. Annie Besant, "Industry under Socialism," in *Fabian Essays in Socialism*, ed. George Bernard Shaw (London: Fabian Society, 1889), 152–53. Besant is better known today as a theosophist, but before taking up that philosophy she was a major contributor to Fabian theory.

7. Norman MacKenzie and Jeane MacKenzie, *The Fabians* (New York: Simon and Schuster, 1977), 108; Graham Wallas, "Property under Socialism," *Fabian Essays in Socialism*, ed. Shaw, 136.

8. Besant, "Industry under Socialism," 153.

9. R. E. Prothero, "Towards Common Sense," *Nineteenth Century* 61 (March 1892): 516; James Stuart, "The London Progressives," *Contemporary Review* 61 (May 1892): 525; Alan McBriar, *Fabian Socialism and English Politics, 1884–1918* (Cambridge: Cambridge University Press, 1962), 197.

10. As the 1891 election approached, the Fabians promoted candidates who were willing to sign on to what they called the London Programme. Sidney Webb to Beatrice Potter, September 1891, *The Letters of Sidney and Beatrice Webb*, ed. N. MacKenzie, 3 vols. (Cambridge: Cambridge University Press, 1978), 1:296.

11. Sidney Webb, *The London Programme* (London: Swan Sonnenschein & Co., 1891), 6.

12. Ibid., 9–11.
13. Ibid., 59–60, 89, 94–96.
14. Ibid., 120–21.
15. Ibid., 74–87.
16. Ibid., 136.
17. Ibid., 53.
18. The issue of water supply provided content in the majority of the well-known *Fabian Essays in Socialism* edited by Shaw, was part of the content of numerous published speeches, including Webb's "Rome: A Sermon in Sociology," *Our Corner* 11 (April 1888), was discussed as an area needing reform in his *Socialism in England* (London: Swan Sonnenschein and Co., 1890), was the first utility Webb proposed to municipalize in *The London Programme*, was a major component of Fabian Tract nos. 8, 10, 21 24, and 26, and was the sole subject of *London's Water Tribute,* Fabian Tract no. 34 (1891).
19. Water's connection to rent, though, has been overlooked by historians. See, for example, David M. Ricci, "Fabian Socialism: A Theory of Rent as Exploitation," *Journal of British Studies* 9 (November 1969): 105–21.
20. G. Bernard Shaw, "The Basis of Socialism," in *Fabian Essays in Socialism,* ed. Shaw, 26.
21. Stewart D. Headlam, *Christian Socialism,* Fabian Tract no. 42 (1892), 12–14.
22. Sydney Olivier, *Capital and Land,* Fabian Tract no. 7 (1893), 5.
23. Sidney Webb, *Figures for Londoners,* Fabian Tract no. 10 (1890), 32.
24. Ibid., 1–2.
25. Headlam, *Christian Socialism,* 13.
26. Sidney Webb, "The Difficulties of Individualism," *Economic Journal* 1 (June 1891): 368–69.
27. Sidney Webb, *London's Water Tribute,* Fabian Tract no. 34 (1890), 3.
28. Webb was especially hopeful about a transformed London's ability to provide services, water in particular, inexpensively. The capitalist, competitive water supply was unnecessarily costly, he argued; the competition of eight companies resulted in eight separate sets of mains, stations, and staff; it also resulted in a contest to provide higher dividends for shareholders, taking away capital from improvements. Webb, *London's Water Tribute,* 1.
29. Webb, *London Programme,* 208.
30. Shaw, "Basis of Socialism," 27.
31. George Bernard Shaw, *A Manifesto,* Fabian Tract no. 2 (1884), 55.
32. Webb, *Figures for Londoners,* 3.
33. Webb, *Facts for Londoners,* 33.
34. Shaw, "Basis of Socialism," 4.
35. Besant, "Industry under Socialism," 65.

36. Wallas, "Organization of Society," 135.

37. Webb, *Facts for Londoners,* 32.

38. Sidney Webb, "Rome: A Sermon in Sociology," *Our Corner* 11 (April 1888): 79.

39. Ibid., 80.

40. Ibid., 84, 60, respectively.

41. Ibid., 82.

42. Webb, *London Programme,* 39.

43. Webb, *Facts for Londoners,* 43.

44. Sidney Webb, "Historic," in *Fabian Essays in Socialism,* ed. Shaw, 50.

45. Ibid., 51.

46. Hubert Bland, "The Outlook," in *Fabian Essays in Socialism,* ed. Shaw, 195.

47. G. Bernard Shaw, "Transition," in *Fabian Essays in Socialism,* ed. Shaw, 170.

48. Webb, "Historic," 35.

49. Sidney Webb, *The Reform of London* (London: Eighty Club, 1892), 12.

50. Shaw, "Transition," 187.

51. Sidney Webb to Beatrice Potter, November 1891, *Letters of Sidney and Beatrice Webb,* 1:328.

52. Webb campaign flier, Passfield Collection: 4/1 London County Council, 1892–February 1907, London School of Economics Archives.

53. A survey of all of the proceedings of the LCC's water committee shows that Webb attended most of the frequent meetings. LCC/MIN series, London Metropolitan Archives.

54. On Farrer's support, see Sidney Webb to Beatrice Webb, May 1892, *Letters of Sidney and Beatrice Webb,* 1:408.

55. See, for example, Webb's speech at the Union, in W. H. Dickinson, T. McKinnon Wood, James Stuart, and Sidney Webb, *Four Speeches on the Present Position of the London Water Supply* (London: London Liberal and Radical Union, 1896).

56. Sidney Webb to Beatrice Webb, June 1899, *Letters of Sidney and Beatrice Webb,* 2:110.

57. Parliament voted on LCC water company purchase bills in 1895, 1896, 1897, 1900, and 1901. The fate of this dogged parliamentary effort is told in chapter 4.

58. The first session of the LCC framed its efforts toward purchasing London's water companies with rhetoric comparable to that of the sanitarians and of Birmingham's water activists. Progressives spoke in the often-heard terms of "scandalous dirty water"; they petitioned to punish recalcitrant companies, and some wished to replace the companies with a "water trust." But the Fabi-

ans provided new theories and tactics to subsequent London councils, especially as the older guard of respectable LCC water committee members retired or moved on.

59. Robert Millward, "Privatisation in Historical Perspective: The UK Water Industry," in *Money, Trade and Payments: Essays in Honour of D. J. Coppock,* ed. David Cobham, Richard Harrington, and George Zis (Manchester: Manchester University Press, 1989), 206.

60. The Fabian attempt to reform the entire water provision system stands outside the story Patrick Joyce tells in *The Rule of Freedom.* In that book, Joyce analyzes the power dynamic behind local governments' projects of urban infrastructure reform in Liberal Victorian Britain. Those projects were as much about passively programming society as liberating it. Technologies for increasing townspeople's freedom to move around, to see and understand the city, to be free from the dangers of crowds and disease—that is, new technological systems for transportation, lighting, and waterworks—were as much about making certain behaviors possible while inhibiting others. They were projects that at once guided people in living a certain "modern" way of life and that quietly reinforced the power of the Liberal regime that undertook such projects. As much as the Fabians insisted that they appealed to the English instinct toward freedom, they were more interested in justice than in the Liberal brand of freedom. Patrick Joyce, *The Rule of Freedom: Liberalism and the Modern City* (London: Verso, 2003).

61. Webb, *Reform of London,* 33.

62. For a survey of the literature on British imperial water projects in India, see Rohan D'Souza, "Water in British India: The Making of a Colonial Hydrology," *History Compass* 4, no. 4 (2006): 621–28. Any work on the history of water control is written in the shadow of Donald Worster's *Rivers of Empire: Water, Aridity, and the Growth of the American West* (New York: Pantheon Books, 1985). He suggests that the society of the western United States was a unique one and that the outcomes generated by water control in the western states were new. There, water schemes led to "the domination of some people over others," as was the case from the Ganges to Euphrates. Ibid., 50. But Worster describes how, because of the evolution of the American economic culture of capitalism, water projects were the tools of new centers of power: influential agriculturalist capitalists on the one hand and state agencies and politicians on the other. Capitalists sought water control in order to amass profits, and state officers and agencies sought it because, in the modern American culture, they viewed their purpose as being to maximize the resources available for conversion into private wealth. Worster has infused sophistication into the literature on water and power, stagnant since the days of Karl August Wittfogel, by showing that the kinds of political and social outcomes generated by water control systems

depended on the particular political and cultural order in which they were constructed.

Chapter 4. From Engineering Modernization to Engineering Collectivization

1. Richard Walker, *The Country in the City: The Greening of the San Francisco Bay Area* (Seattle: University of Washington Press, 2008), 6 and elsewhere; William Cronon, *Nature's Metropolis: Chicago and the Great West* (New York: Norton, 1989), 7.

2. For a description of this concept, see Jared Orsi, *Hazardous Metropolis: Flooding and Urban Ecology in Los Angeles* (Berkeley and Los Angeles: University of California Press, 2004).

3. London County Council, *Report on Available Sources of Water Supply for London, 1894* (London: LCC, 1894), 6.

4. [Matthew White Ridley] *Select Committee on Metropolitan Water Supply, 1891* (London: HMSO, 1891).

5. Orsi, *Hazardous Metropolis*, 167.

6. This general conception of the first LCC's unwritten program owes much to John Davis, *Reforming London: The London Government Problem, 1855–1900* (Oxford: Clarendon Press, 1988), 122–25.

7. Joseph Chamberlain, quoted in Asa Briggs, *Victorian Cities* (New York: Harper and Row, 1965), 239.

8. J. L. Garvin, *The Life of Joseph Chamberlain*, 6 vols. (London: Macmillan, 1932–69), 1:192–93.

9. Davis, *Reforming London*, 133.

10. Stephen Inwood, *A History of London* (New York: Carroll & Graf, 1998), 441; Davis, *Reforming London*, 132.

11. Frederic Harrison, "Lord Rosebery and the London County Council," *Nineteenth Century* 59 (June 1890): 1030; Inwood, *History of London*, 441.

12. Susan D. Pennybacker, *A Vision for London, 1889–1914: Labour, Everyday Life and the LCC Experiment* (New York: Routledge, 1995), 30.

13. Davis, *Reforming London*, 136.

14. Ibid., 133–34.

15. For evidence that the Progressives took their policy from *The London Programme*, see R. E. Prothero, "Towards Common Sense," *Nineteenth Century* 31 (April 1892): 516; and [London MP] James Stuart, "The London Progressives," *Contemporary Review* 61 (May 1892): 525. See also A. M. McBriar, *Fabian Socialism and English Politics, 1884–1918* (Cambridge: Cambridge University Press, 1962), 197.

16. Stuart, "London Progressives," 526.

17. McBriar, *Fabian Socialism*, 197.

18. See the discussion in Briggs, *Victorian Cities*, 337. The Progressive coun-

cilor John Burns used the term *commune*. John Burns, "Towards a Commune," *Nineteenth Century* 31 (April 1892): 498–514. For the basis of the Progressives' campaign, see Burns, "Towards a Commune," 500–502. Portions of the campaign address by the Progressive John Benn are reprinted in A. G. Gardiner, *John Benn and the Progressive Movement* (London: Ernest Benn, 1925), 152; see also ibid., 155. On the general nature of the Progressives' common platform for the second LCC elections, see Briggs, *Victorian Cities*, 330, 339.

19. Burns, "Towards a Commune," 501. See also "Let London Live," *Nineteenth Century* 31 (1892): 675; and Gardiner, *John Benn*, 151.

20. Stuart, "London Progressives," 521.

21. These were F. C. Baum, Will Crooks, Fred Henderson, Frank Smith, Will Steadman, Sidney Webb, and Ben Tillet and John Burns, respectively. Paul Thompson, *Socialists, Liberals and Labour: The Struggle for London, 1885–1914* (Toronto: University of Toronto Press, 1967), 146.

22. On this transition, see Davis, *Reforming London*, 119, 125.

23. McBriar, *Fabian Socialism*, 197; Briggs, *Victorian Cities*, 377; Hope Costley-White, *Lord Dickinson of Painswick: A Memoir*, also titled *Willoughby Hyett Dickinson, 1859–1943: A Memoir* (Gloucester: privately printed by John Bellows Ltd., 1965), 25; Lord Hobhouse, "The London County Council and Its Assailants," *Contemporary Review* 61 (April 1892): 340.

24. Sir John Lubbock, Sir T. H. Farrer, and the alderman Lord Hobhouse, to name a few. Davis, *Reforming London*, 153.

25. Davis, *Reforming London*, 153–54; Briggs, *Victorian Cities*, 337.

26. Gardiner, *John Benn*, 154.

27. Wemyss quoted in Ken Young, *Local Politics and the Rise of the Party: The London Municipal Society and the Conservative Intervention in Local Elections, 1894–1963* (Leicester: Leicester University Press, 1975), 52.

28. C. A. Whitmore, "Conservatives and the London County Council," *National Review* 21 (April 1893): 183.

29. The London Municipal Society's platform is reprinted in the *Times*, 7 July 1894, 14.

30. Davis, *Reforming London*, 154.

31. The full text of the speech is printed in the *Times*, 8 November 1894, 4.

32. Young, *Local Politics*, 59.

33. Whitmore, "Conservatives and the London County Council," 184.

34. John Frederic Bateman, *On the Supply of Water to London from the Sources of the River Severn* (London: Vacher and Sons, 1865).

35. London County Council, Engineer's Report on Water Supply, 1890, 8 and 29 October 1890 (London: LCC, 1890), 16, 18.

36. London County Council, *Engineer's Report on the London Water Supply from the Thames and Lea, 1 September 1891* (London: LCC, 1891), 6.

37. Royal Commission on Metropolitan Water Supply, *1893 Report* (London: HMS, 1893), esp. 71.

38. Ibid., 72.

39. "We acknowledge the great importance to London of the evidence collected by the Royal Commission," reported the water committee to the full LCC, "but we do not consider that the main suggestions of the Report are likely to afford a final and satisfactory settlement of the London Water Question." Draft of Proposed Report from [the LCC Water Committee regarding the Royal Commission on Water Supply report], 2, LCC/MIN/12,271, London Metropolitan Archives.

40. Municipal water enterprises in 1907 in England and Wales yielded a 5 percent profit on cities' expenditures, most of which was employed in lowering rates. Douglas Knoop, *Principles and Methods of Municipal Trading* (London: Macmillan and Co., 1912), 311. Sidney Webb expected municipalization of water and other services to increase debt in the short term but result in real savings for consumers overall. Sidney Webb, *Facts for Londoners,* Fabian Tract no. 8 (1889), 54.

41. Asok Kumar Mukhopadhyay, *Politics of Water Supply: The Case of Victorian London* (Calcutta: World Press, 1981), 85.

42. Salisbury speech before Moderates and London MPs, transcribed in *Times,* 26 March 1895, 10.

43. *Times,* 4 September 1895, 7; Memorandum from the chief of the fire brigade to the Water Committee, 25 September 1895, 1, LCC/MIN/12,271. The high temperature that day was 79 degrees.

44. *Times,* 29 July 1895, 11.

45. *Times,* 1 July 1895, 7. The figure of 5.83 inches indicates the level below the mean for the previous twenty-five years.

46. *Times,* 30 September 1895, 7. London still suffered a deficit of 4.88 inches.

47. On September 24, the high temperature in London was 87 degrees. *Times,* 25 September 1895, 4. Average high temperatures calculated from those reported daily in the *Times* during September 1895.

48. Londoner Robert Ramsey, diary of June and July 1895 (pages unnumbered), F/RMY/21, London Metropolitan Archives. He begins each diary entry with a general weather report.

49. *Times,* 6 August 1895, 10; A. R. Binnie, Report on the East London Water Company, Alleged Failure of Supply, 10 March 1896, 1, LCC/MIN/12,272, London Metropolitan Archives.

50. Letter from an unnamed Hackney resident to the *Times,* 23 July 1895, 12.

51. Binnie, Report on the East London Water Company, Alleged Failure of Supply, 1.

52. W. H. Dickinson, "Water Supply of Greater London," part 19, *Engineer* 186 (November 1948): 483.

53. Mukhopadhyay, *Politics of Water Supply*, 101–2, 161n3.

54. Mike Hulme and Elaine Barrow, *Climates of the British Isles: Present, Past and Future* (London: Routledge, 1997), 406, 409–10.

55. "The southern and eastern parts of England . . . have again been sufferers from drought," reported the *Times* at the beginning of September; London had seen less than half of its usual August rainfall. *Times,* 2 September 1898, 8. On these droughts and the social and political consequences in the large area of West Ham in northeast London, see Jim Clifford, "A Wetland Suburb on the Edge of London: A Social and Environmental History of West Ham and the River Lea, 1855–1914 (PhD diss., York University, Toronto, 2011).

56. *Times,* 17 September 1898, 11; *Daily Mail,* 10 September 1898, 3.

57. Waterworks correspondence [belonging to the Local Government Board], August to September 1898, PRO-MH-29-33, National Archives.

58. *Star,* 24 August 1895, 3.

59. *Star,* 25 August 1895, 3; *Daily Mail,* 10 September 1898, 3.

60. Waterworks Correspondence [belonging to the Local Government Board], August to September 1898.

61. Letter of twelve doctors and medical professionals, quoted in the *Star,* 24 August 1898, 3.

62. Balian, member for St. George's-in-the-East, quoted in the *Times,* 7 September 1898, 9.

63. Waterworks Correspondence [belonging to the Local Government Board], August to September 1898.

64. Waterworks Correspondence [belonging to the Local Government Board], October to December 1898, PRO MH-29-34, National Archives.

65. W. H. Dickinson, quoted in the *Echo,* 18 October 1898, 3.

66. Quoted in the *Times,* 5 October 1898, 10. Stuart was also a financial backer and board member of the *Star.* Norman Mackenzie, ed., *The Letters of Sidney and Beatrice Webb,* 3 vols. (Cambridge: Cambridge University Press, 1978), 1:129.

67. Both quoted in the *Times,* 12 October 1898, 10.

68. W. H. Dickinson, "Private and Confidential Memorandum by the Chairman of the Water Committee with reference to proposals for legislation in the Session of 1899," 15 October 1898, 1, LCC/MIN/12,274, London Metropolitan Archives. This is a strategy memo to allies on the LCC's water committee in which he proposed action for the coming session.

69. Bethnal Green, 129,162; Hackney, 213,044; Mile-end Old Town, 111,060; St. George's-in-the-East, 47,506; Limehouse, 58,306; Poplar, 169,267; Whitechapel, 77,757. *Times,* 28 September 1898, 6.

70. Dickinson, "Private and Confidential Memorandum," October 15, 1898, 1.

71. The *Star* asked in one heading, "Why Not Purchase Water?" The newspaper quoted a Thames Conservancy official who said it was technically feasible. *Star,* 27 August 1898, 3.

72. See various letters of refusal from August and September 1898 in Waterworks Correspondence [belonging to the Local Government Board], August to September 1898.

73. Dickinson, "Private and Confidential Memorandum," 15 October 1898, 1.

74. The LCC engineer had reported to the council that interconnecting the existing companies' systems of distribution was feasible. London County Council, *Report on Available Sources of Water Supply for London, 1894,* 12. In a confidential memorandum of 15 October 1898, Dickinson wrote that the specific issue of East London failures could be solved through LCC interconnection, while he believed the local failures suggested total London-wide failure of the water supplies in ten to fifteen years. Dickinson, "Private and Confidential Memorandum," 15 October 1898, 1.

75. Dickinson, "Private and Confidential Memorandum," 15 October 1898, 2.

76. *Journal of Gas Lighting, Water Supply & Sanitary Improvement* 72 (30 August 1898): 424, 475. The authors of one study find that water consumer organization was self-organized and not mere political theater. See Vanessa Taylor, Heather Chappells, Will Medd, and Frank Trentmann, "Drought Is Normal: The Socio-Technical Evolution of Drought and Water Demand in England and Wales, 1893–2006," *Journal of Historical Geography* 35, no. 3 (2009): 582.

77. Though it is difficult to determine whether or not Progressives or Fabians took part in demonstrations, newspaper reports tended to convey who chaired or chiefly contributed to public meetings. Trentmann and Taylor suggest that in at least one instance they find evidence of LCC members following, not leading, grass-roots protesters. Vanessa Taylor and Frank Trentmann, "Liquid Politics: Water and the Politics of Everyday Life in the Modern City," *Past and Present* 211 (May 2011): 222.

78. Letter from the "Meeting of the Inhabitants of Camberwell," 4 December 1896, LCC/MIN/12,272, London Metropolitan Archives.

79. Waterworks Correspondence [belonging to the Local Government Board], August to September 1898.

80. Letters of vestries to the LCC, 13 August 1898, 4 October 1898, and 22 September 1898, respectively, LCC/MIN/12,274.

81. Letters from the Whitechapel Board and the Limehouse District Board to the LCC, 31 July 1896 and 16 September 1896, respectively, LCC/MIN/12,272.

82. Resolutions of Hackney, 6 November 1898, and St. Mary, Stratford Bow, 4 October 1898, LCC/MIN/12,274.

83. At least ten such resolutions of 1898 are recorded in PRO-MH-29-33, National Archives, and LCC/MIN/12,274.

84. Davis counts the LCC's fight against the companies among the factors that earned the Progressives a great majority in East London constituencies. Davis, *Reforming London*, 232–33.

85. Lord Onslow, speaking at the Junior Constitutional Club, quoted in the *Journal of Gas Lighting, Water Supply, & Sanitary Improvement* 72 (6 December 1898): 1309.

86. See the description of how the urban ecosystem in Los Angeles made it particularly susceptible to flooding in Orsi, *Hazardous Metropolis*.

87. London County Council, Minutes of Proceedings, 19 March 1889, LCC/MIN/6257.

88. London County Council, Minutes of Proceedings, 15 May 1889, 288–90, LCC/MIN/6257.

89. London County Council, *Annual Report of the Proceedings of the Council, Year ending 1891* (London: LCC, 1891), 83–84; London County Council, *Engineer's Report on Water Supply, 1890, 8 and 29 October 1890*.

90. Binnie acknowledged his teacher Bateman by calling himself Bateman's "old pupil" in Alexander Binnie, *Lectures on Water Supply, Rainfall, Reservoirs, Conduits & Distribution*, 2nd ed. (Chatham: W. & J. Mackay & Co., 1887), iii. See also Report from the Select Committee on London Water Bills (Lambeth and Southwark and Vauxhall Companies transfer), 1895, Minutes of Evidence, (London: HMSO, 1895)182.

91. *Times*, 19 May 1917, 8.

92. London County Council, *Report on Available Sources of Water Supply for London*, 3.

93. Binnie in London County Council, *Engineer's Report on Water Supply, 1890, 8 and 29 October 1890*, 16–18.

94. Ibid., 18.

95. Alexander Binnie, letter to the Special Committee on Water Supply and Markets, 15 December 1890, 1, LCC/MIN/12,267, London Metropolitan Archives. "This I hope will result in all the data . . . being placed on a much firmer basis than hitherto as the information to be obtained on the 6-inch sheets did not exist at the time that the projects were originally brought before the Royal Commission in 1869," he wrote. Ibid. Binnie considered other options for an improved source of water supply, but none won his favor. His concern about the demands of an expanding metropolitan population on one hand and a related increase in run-off pollution into the Thames on the other made him anxious to secure a water supply that was as abundant and as far away from population centers as possible. He rejected a plan to dig a series of wells in Kent as inadequate, and a scheme of building reservoirs up-

river in the Thames valley he deemed too perilous for future generations of Londoners. Shortly after sending his team of investigators to Wales, Binnie reported serious concerns about the danger of using Thames water; he wrote that even if the Thames "can yield, even after the construction of costly works, a quantity sufficient only for a comparatively short period, it may be deemed prudent to concentrate further expenditure on some new source of supply, rather than on an attempt to increase the supply from the Thames and Lea valleys." Ibid. In sum, Binnie was focused on Wales from the earliest point of his investigations; when he did write of alternatives, he discussed only their shortcomings. See W. Whitaker and A. H. Green, Geologists' Report, 27 April 1891, 25, LCC/MIN/12,267. See also London County Council, *Engineer's Report on the London Water Supply from the Thames and Lea, 1891, 1 September 1891*, 5–6.

96. J. H. Balfour Browne, *Water Supply* (London: MacMillan and Co., 1880), 15.

97. London County Council, *Report on Available Sources of Water Supply for London*, 13.

98. Binnie, letter to the Special Committee on Water Supply and Markets, 15 December 1890, 3.

99. Ibid., 3–4.

100. Welsh towns, too, were reaching into their own hinterlands and constructing reservoirs and aqueducts. The LCC's water committee noted that, in 1891, Swansea had obtained authority to take water from a tributary of the Usk, a river on which London also had an eye. *Report of the Water Committee of the London County Council on Supplementary Supply, 25 November 1895 and 21 April 1896* (London: LCC, 1896), 119.

101. *Report of the Special Committee on Water Supply of the London County Council, 1891* (London: LCC, 1891).

102. LCC, Water Committee Minutes, 22 February 1892, LCC/MIN/12,266, London Metropolitan Archives.

103. LCC, Special Water Committee Memorandum, 18 March 1892, "Special Water Committee" (Presented) Papers, 1892, LCC/MIN/12,268, London Metropolitan Archives.

104. Chair of the Birmingham Water Committee, speech quoted by Sir John Lubbock, London county councilor and MP, in *Parliamentary Debates*, Commons, 4th ser., vol. 2 (8 March 1892), col. 284.

105. Sir Hussey Vivian, *Parliamentary Debates*, Commons, 4th ser., vol. 2 (8 March 1892), col. 288.

106. The LCC petitioned against the Birmingham water bill in the House of Lords, "in view of the possibility . . . that the present sources of supply cannot be extended . . . and in view also of the fact that the possible fresh areas

of supply are limited." London County Council, Water Committee Minutes, 2 June 1892, 72, LCC/MIN/12,266.

107. George Shaw-Lefevre, "The London Water Supply," *Nineteenth Century* 67 (1898): 980. In fact, the specter of the Welsh project frightened company shareholders even before the LCC proposed its plan. At a shareholder meeting in 1880, F. Banbury, who owned shares in the East London Company, said that "nearly all the papers unanimously say that these companies have no monopolies whatever, and that there is nothing to prevent a new scheme being authorized for the purpose of bringing water from Mid Wales at a cost of eight millions; the engineers' estimates for which are prepared. . . . To my mind [such prospects] comprise so entirely the very points on which these negotiations in the future will turn." F. Banbury, quoted in "Minutes of Proceedings at an Extraordinary General meeting of the Proprietors of the East London Waterworks, held at the Cannon Street Hotel, on Monday, March 22, 1880," 12, Stanford Sanitary Reform Collection, Special Collections, Stanford University Libraries, Stanford, CA.

108. Fabian Society [Sidney Webb], *London's Water Tribute,* Fabian Tract no. 34 (1891), 1.

109. Fabian Society [Sidney Webb], *Facts for Londoners,* Fabian Tract no. 8 (1889), 32.

110. Sidney Webb, *Three Years' Work on the London County Council: A Letter to the Electors of Deptford* (London: T. May, 1895), 13–14.

111. Ibid., 24.

112. Ibid., 25.

113. G. M. Binnie, *Early Victorian Water Engineers* (London: Telford, 1981), 130–31.

114. H. C. Richards and W. H. C. Payne, *London Water Supply,* 2nd ed. (London: P. S. King and Son, 1899), 228; Alfred Lass, *London Water Supply, Analysis of the Accounts of the Metropolitan Water Companies for the Year Ended December 31/March 31 1896* (London: Walter King, 1896), 6.

115. Sidney Webb, *The London Programme* (London: Swan Sonnenschein & Co., 1891), 208–9. A few years earlier, in *Facts for Londoners,* too, Webb had called on the LCC "to construct a new supply," and "for constant supply [to] be made universal." Webb, *Facts for Londoners,* 33.

116. Webb, *London Programme,* 208–9.

117. Webb, *Facts for Londoners,* 43.

118. Alexander Binnie, letter to the LCC Water Committee, 17 November 1891, LCC/MIN/12,267; Engineer's Expense Account, 18 October 1892, LCC/MIN/12,268.

119. London County Council, *Report on Available Sources of Water Supply for London,* 3.

120. Ibid., 12.

121. For the plan in detail, including water analysis and other particulars, see London County Council, *Report on Available Sources of Water Supply for London,* 14–17. Also see the final report presenting the Welsh scheme: *Report of the Water Committee of the London County Council on Supplementary Supply, 25 November 1895 and 21 April 1896* (reproduced in full in the appendices of the [Llandaff] Royal Commission on Water Supply, 1899, 116–23), 121.

122. London County Council, *Report on Available Sources of Water Supply for London,* 4.

123. Ibid., 5.

124. *Red Dragon: The National Magazine of Wales* 6 (1884): 357–58.

125. *Current Literature* 21 (April 1897): 364.

126. Great Britain, Rivers Pollution Commission, 1868, *Fourth Report of the Commissioners,* vol. 1, *Reports and Plans* (1872), 11.

127. For the self-purification debate, see Christopher Hamlin, *A Science of Impurity: Water Analysis in Nineteenth-Century Britain* (Berkeley and Los Angeles: University of California Press, 1990), 78, 174, and elsewhere.

128. *Practitioner* 16 (May 1876): 415–16; Royal Society of Arts, *National Water Supply: Notes on Previous Inquiries* (London: George Bell, 1878), 100. Also see, generally, the reports of the Rivers Pollution Commission of 1868.

129. London County Council, *Report on Available Sources of Water Supply for London,* 10.

130. Ibid., 6.

131. Ibid., 9.

132. W. H. Dickinson, *Water for London,* London Reform Union Pamphlet no. 78 (1898), 16.

133. Ibid., 14.

134. *Daily Chronicle,* 1 October 1895, 8. See, as another example, *The Sword and Shield* (London), 21 June 1890, repeating Webb's call for an alternative supply from *Facts for Londoners* almost verbatim.

135. North London Property Owners' Association, letter to the London County Council, 23 July 1895; and Camberwell Vestry, letter to London County Council, 2 August 1895, both in LCC/MIN/12,271. See also Resolution of the Wandsworth Vestry, 28 January 1899, LCC/MIN/12,275, London Metropolitan Archives; and W. Howles, letter to the LCC Water Committee, 12 February 1895, LCC/MIN/12,271.

136. B. F. C. Costelloe, "London v. the Water Companies," *Contemporary Review* 67 (1895): 812–13.

137. W. H. Dickinson, "The Water Supply of London," *Contemporary Review* 69 (1897): 244–45.

138. Ibid., 245.

139. C. M. Knowles, *Municipal Water,* Fabian Tract no. 81 (1898), 1–3. The

London Reform Union echoed the admonishing line, writing that "every other large town in Great Britain has given up drinking polluted river water." London Reform Union, *Why Is Your Water Rate So High?*, London Reform Union Pamphlet no. 48 (1895), 4.

140. Either because of frequent calls to abandon the apparently failing rivers, or because Parliament had permitted a number of other cities to draw a water supply from Wales, or simply because the idea had been circulated for decades, even in disinterested quarters there grew some sense that, eventually, London would get a water supply from Wales. See, for example, *Saturday Review,* 27 August 1898, 262; and Joseph Pennel, "The Welsh 'Corice,'" *Contemporary Review* (April 1899): 522–33.

141. Dickinson, "Water Supply of London," 245.

142. Ibid., 243.

143. Shaw-Lefevre, "London Water Supply," 988.

144. According to a book written on the water controversy while the royal commission was meeting, the government appointed the commission in response to "continued agitation against the companies." Richards and Payne, *London Water Supply,* 200.

145. [Henry Chaplin], Memorandum upon the Metropolitan Water Supply, 6 January 1897, 11, CAB/37/44/1, National Archives.

146. Ibid., 9.

147. House of Commons, *Sessional Papers,* 1900, vol. 38, Water Supply, part 1, vii.

148. Dickinson, *Water for London,* 6–7.

149. *Report of the Water Committee of the London County Council on Supplementary Supply, 25 November 1895 and 21 April 1896,* 122.

150. House of Commons, *Sessional Papers,* 1900, vol. 38, Water Supply, part 3, 212–13.

151. House of Commons, *Accounts and Papers,* 1900, vol. 50, "Navy Estimates," part 1, 2.

152. Edward Pember, QC, was counsel for the Lambeth, East London, Grand Junction, and West Middlesex Companies. House of Commons, *Sessional Papers,* 1900, vol. 38, Water Supply, part 1, 480.

153. *Times,* 6 April 1911, 11.

154. House of Commons, *Sessional Papers,* 1900, vol. 38, Water Supply, part 1, 348.

155. Ibid., 353.

156. H. Wilkins, secretary to the Lambeth Company, House of Commons, *Sessional Papers,* 1900, vol. 38, Water Supply, part 1, 476.

157. Walter Hunter, speech before the Society of Arts, April 1899, in "London Water Supply," *Journal of the Society of Arts* 47 (21 April 1899): 486–87.

158. The Staines supply was shared out among half the companies through

the Staines Reservoir Joint Committee, created in 1896. House of Commons, *Sessional Papers,* 1900, vol. 38, Water Supply, part 1, 67.

159. House of Commons, *Sessional Papers,* 1900, vol. 38, Water Supply, part 2, 653.

160. Ibid., 645.

161. Ibid., part 3, 234.

162. See House of Commons, *Sessional Papers,* 1900, vol. 38, Water Supply, part 2, 126.

163. Quoted in ibid., part 1, 467.

164. Ibid., 507. By this point in time, Binnie had been knighted Sir Alexander for having completed the celebrated Blackwall pedestrian tunnel under the Thames in 1897.

165. Quoted in House of Commons, *Sessional Papers,* 1900, vol. 38, Water Supply, part 1, 466.

166. Ibid., part 3, 252–53.

167. Ibid., part 2, 130. Surprisingly, peat was considered as pernicious as sewage.

168. Edward Pember testimony in House of Commons, *Sessional Papers,* 1900, vol. 38, Water Supply, part 1, 507.

169. James Dewar, quoted in House of Commons, *Sessional Papers,* 1900, vol. 38, Water Supply, part 2, 268. Professor Dewar was employed by the Local Government Board to make monthly chemical and bacteriological examinations of the water companies' supplies.

170. Hunter speech, in "London Water Supply," *Journal of the Society of Arts,* 493.

171. Ibid., 494.

172. House of Commons, *Sessional Papers,* 1900, vol. 38, Water Supply, part 1, 49.

173. Ibid.

174. Frederick Dixon-Hartland, *Parliamentary Debates,* Commons, 4th ser., vol. 68 (21 March 1899), col. 1539.

175. C. A. Whitmore, *Parliamentary Debates,* Commons, 4th ser., vol. 68 (21 March 1899), col. 1538.

176. James Stuart, *Parliamentary Debates,* Commons, 4th ser., vol. 68 (21 March 1899), col. 1532.

177. Frederick Dixon-Hartland, *Parliamentary Debates,* Commons, 4th ser., vol. 68 (21 March 1899), col. 1532, 1539.

178. One scholar dates the end of Lloyd George's Welsh nationalism to this very period, writing, "After this disheartening setback [the failure to officially change the Welsh caucus in Commons into an autonomous party], Lloyd George's preoccupation with Welsh nationalism ended, though it is too much to say, as did one Welsh Liberal politician, that thereafter he became 'a below

the gangway English radical—nothing more.'" Don M. Cregier, *Bounder from Wales: Lloyd George's Career before the First World War* (Columbia: University of Missouri Press, 1976), 63.

179. A. C. Humphreys-Owen, *Parliamentary Debates,* Commons, 4th ser., vol. 68 (21 March 1899), col. 1541. Humphreys-Owen did not elaborate on the "many reasons" Wales should be glad to welcome the LCC, but reservoir projects certainly injected money into local economies, often brought along infrastructure in the form of rail lines, and left behind lakes that drew visitors.

180. James Stuart, *Parliamentary Debates,* Commons, 4th ser., vol. 68 (21 March 1899), col. 1543.

181. Ibid., col. 1535.

182. Henry Chaplin, *Parliamentary Debates,* Commons, 4th ser., vol. 68 (21 March 1899), col. 1544.

183. Ibid., col. 1546.

184. *Daily Chronicle,* 22 March 1899, 3.

185. Quoted in *Times,* 29 March 1899, 10.

186. Quoted in *Daily Chronicle,* 22 March 1899, 3.

187. Ibid. Maidstone, southeast of London, suffered a devastating typhoid outbreak in 1899.

188. At the same meeting, the councilor E. A. Cornwall still held out hope for a plan, vowing that "it would be the duty of the Council year after year to present schemes for dealing with the London water question until it obtained the adoption of its policy by Parliament." Quoted in *Times,* 29 March 1899, 10.

189. Alexander Binnie, *Report on the Council's Water Bills, 16 June 1899* (London: LCC, 1899), 1.

190. London County Council, Water Committee Minutes, 4 May 1900, LCC/MIN/12,266.

191. W. C. Steadman, *Parliamentary Debates,* Commons, 4th ser., vol. 69 (18 April 1899), col. 1494–95.

192. Quoted in House of Commons, *Sessional Papers,* 1900, vol. 38, Water Supply, part 1, 344–45.

193. James C. Scott, *Seeing Like a State: How Certain Schemes to Improve the Human Condition Have Failed* (New Haven: Yale University Press, 1998); Erik Swyngedouw, "Modernity and Hybridity: Nature, Regeneracionismo, and the Production of the Spanish Waterscape, 1890–1930," *Annals of the Association of American Geographers* 89 (September 1999): 443–65.

Chapter 5. An Alternative Vision of the Modern City, an Alternative Government of Water

1. *Times,* 23 January 1899, 6.

2. Quoted in *Times,* 25 January 1899, 11.

3. Ken Young, *Local Politics and the Rise of Party* (Leicester: Leicester University Press, 1975), 86.

4. The committee members were Viscount Hampden, Viscount Peel, Lord Rothschilde, Sir Leonard Lyell, Grant Lawson, Lewis Fry, L. T. Hobhouse, and, serving as chair, the Earl of Crewe.

5. *Report from the Joint Select Committee on Municipal Trading, 1900* (London: HMSO, 1900), 335.

6. Ibid., 170.

7. Ibid., 171, 177.

8. Ibid., 126, 295, 277–78.

9. Ibid., 289, 326.

10. Quoted in ibid., 127.

11. *Report from the Joint Select Committee on Municipal Trading, 1900,* 127.

12. Asok Kumar Mukhopadhyay, *Politics of Water Supply: The Case of Victorian London* (Calcutta: World Press, 1981), 104.

13. Vestry of St. Mary, Battersea, "Summary of replies received from the Metropolitan Vestries and District Boards to the Circular from the Vestry upon the question of the London Water Supply," 30 November 1898, LCC/MIN/12,274, London Metropolitan Archives.

14. Quoted in *Times,* 12 July 1899, 3.

15. London County Council, *Report of the London County Council for the year 1900–1901,* (London: LCC, 1904), 203.

16. Lord James of Hereford, Confidential Memorandum upon the Metropolitan Water Supply, 6 January 1897, 11, CAB/37/44/1, National Archives.

17. Henry Chaplin, Confidential Memorandum on London Water, 2 December 1898, 5, CAB/37/48/90, National Archives; Lord James, Confidential Memorandum upon the Metropolitan Water Supply, 6 January 1897, 11. In 1900, Walter Long, president of the Local Government Board, was still holding out hope that the Progressives would soon lose control of the LCC, thus changing the picture of the water question. W. H. Long, Confidential Memorandum to the Cabinet on London Water Supply from the President of the Local Government Board, 17 November 1900, CAB/37/53/74, National Archives.

18. Chaplin, Confidential Memorandum on London Water, 2 December 1898, 5.

19. Full text of Salisbury's speech reprinted in *Daily Chronicle,* 17 November 1897, 6.

20. John Davis, *Reforming London: The London Government Problem, 1855–1900* (Oxford: Clarendon Press, 1988), 237.

21. On the case of the government reform committee in general, see ibid., 236.

22. W. J. Collins, "The London Government Bill," *Contemporary Review* 75 (April 1899): 520, 521.

23. In council debate, W. H. Dickinson, for example, called the devolution effort "dangerous" to poorer parts of London that were separated from the interests and power of wealthier parts. LCC debate transcribed in *Daily Chronicle,* 8 March 1899, 8.

24. *Star,* 17 November 1897, 1; *Daily Chronicle,* 25 January 1899, 7.

25. "National Institutions and Popular Demands," *Blackwood's Edinburgh Magazine* 165 (February 1899): 456, 457.

26. *City Press,* 26 October 1898, 4; 26 January 1898, 4–5.

27. *Times,* 10 May 1899, 11.

28. "Progressivism objects to the devolution of powers in Metropolitan self-government, because, for one reason, it would put a stop to the municipalization of the water and gas supply, the tramways, and many other services," opined the editors of an industry publication. *Journal of Gas Lighting, Water Supply, & Sanitary Improvement* 72 (30 August 1898): 473–74. "We have nothing to do here with . . . party politics," they continued, "but with the prospects of Municipalism in London we certainly are very much concerned. . . . The London County Council . . . wish for a large extension of the powers of the Council. . . . It will try to run rough-shod over a Water Company with only law and justice on their side." Ibid., 72 (6 December 1898): 1280.

29. Davis, *Reforming London,* 247; Ken Young, "The Politics of Local Government, 1880–1899," *Public Administration* 51 (spring 1973): 105.

30. *Municipal Journal* quoted in Young, "Politics of Local Government," 106.

31. Anonymous Marylebone vestryman quoted in *Times,* 2 September 1896, 5.

32. Onslow, letter to Salisbury, 14 February 1897, cited in Davis, *Reforming London,* 208.

33. G. Shaw-Lefevre, "The London Water Supply," *Nineteenth Century* 44 (December 1898): 982.

34. House of Commons, *Sessional Papers,* 1900, vol. 38, Water Supply, part 1, items 90–131.

35. Ibid., item 147.

36. W. H. Dickinson, T. McKinnon Wood, James Stuart, and Sidney Webb, *Four Speeches on the Present Position of the London Water Supply* (London: London Liberal and Radical Union, 1896), 19, 26.

37. See, for example, the communications and memoranda of the Surrey County Council, which concluded that "no change ought to be made from the present system which does not give the control to Surrey of its own sources of water, and the supply thereof [and] the absolute assurance that Surrey will not have to contribute to the cost of any fresh supply for the County of London." Surrey County Council, memorandum to both Houses of Parliament, 28 April 1896, #CC28/70, Surrey History Centre. See also Mukhopadhyay, *Politics of Water Supply,* 94–95, 98.

38. W. H. Long, president of the Local Government Board, confidential memorandum to the Cabinet, 15 July 1901, 2–3, CAB/37/57/69, National Archives.

39. In 1901, there was still pressure emanating from some second-tier authorities for the government to stop obstructing the LCC purchase efforts. "This [London County] Council," resolved the council of the new metropolitan borough of Woolwich in May 1901, "recognising that Birmingham, Manchester, Liverpool, Glasgow, and other large Municipalities enjoy the right to control their own Water Supply, heartily support the claim of a similar right for the people of London, and indignantly protest against the action of His Majesty's Government in opposing and defeating the London County Council's Bill for acquiring the Undertakings of the Water Companies." A public meeting at the Wandsworth town hall in the same month resolved that "this meeting of ratepayers . . . emphatically protest against the action of His Majesty's Government in repeatedly rejecting the Water Purchase Bills of the London County Council. It also affirms its opinion that, in the best interests of the people of London, the County Council should have entire control of the water supply." Wandsworth Town Hall Public Meeting Resolution, 13 May 1901, LCC/MIN/12,276, London Metropolitan Archives.

40. According to the Unionist MP for Shoreditch and Hoxton Hay, "the fact that the London County Council elections upon the last occasion gave such an enormous majority for the Radicals was entirely due to a feeling throughout London that His Majesty's Government were not likely to keep their pledge in regard to the water question and to outrageous regulations of the Water Companies." *Hansard Parliamentary Debates,* 4th ser., vol. 103 (1902), cols. 1372–73. The "regulations" were a proposed requirement that all customers have cisterns on their premises for the storage of water in case of outages. The government waited until 1902 to bring its bill because it hoped, in 1901, that the LCC would change hands, it wanted its new municipal borough councils to mature, and it wanted to give the water companies as much notice as possible of its intentions out of respect for private enterprise. W. H. Long, Confidential Memorandum to the Cabinet on London Water Supply from the President of the Local Government Board, 17 November 1900, 2, CAB/37/53/74, National Archives. See also Mukhopadhyay, *Politics of Water Supply,* 111.

41. Westminster asked for and received a special charter symbolically reconfirming its ancient status as a city in fall 1899, after the passage of the London Government Act.

42. These board members were Sir Edward Fry, Sir Hugh Owen, and Sir John Wolfe Barry, respectively.

43. Walter Long (Viscount Long of Wraxall), *Memories* (London: Hutchinson and Co., 1923), 137. The LCC's supporters later alleged that there was not much interest among the metropolitan boroughs for taking a seat on the water

board, they provided proof that several were plainly against it, and they called on Long to offer proof of strong borough interest—which he could not manage.

44. Long, *Memories,* 136.

45. Scott Edward Roney, "Trial and Error in the Pursuit of Public Health: Leicester, 1849–1891" (PhD diss., University of Tennessee, Knoxville, 2002), 109.

46. Young, *Local Politics and the Rise of Party,* 79; Davis, *Reforming London,* 236.

47. Between 1861 and 1881, the proportion of towns operating a municipal water supply increased from 41 to 80 percent. John Hassan, "The Growth and Impact of the British Water Industry in the Nineteenth Century," *Economic History Review* 38 (November 1985): 535. For a survey of towns with municipalized water supplies and the proportion of those operated by sole town or borough councils and those operated by joint authorities, see Arthur Silverstone, *The Purchase of Gas and Water Works, with the Latest Statistics of Municipal Gas and Water Supply* (London: Crosby Lockwood, 1881), 88–125; Rebe Taylor, *Rochdale Retrospect* (Rochdale: Rochdale Borough Council, 1955), 127–88; Local Government Board, *Urban Water Supply: Return showing the means by which drinkable water is supplied to every urban sanitary district in England, Wales, 3 July 1879* (London: HMSO, 1879), 2–143.

48. Silverstone, *Purchase of Gas and Water Works,* 96–97; Local Government Board. *Urban Water Supply,* 86.

49. David Lewis, *Edinburgh Water Supply: A Sketch of Its History Past and Present* (Edinburgh: Andrew Elliot, 1908), 151; Silverstone, *Purchase of Gas and Water Works,* 91–92. Leigh and Hindley also formed a joint committee for this purpose. Ibid., 112.

50. Such was the case in Belfast. See Silverstone, *Purchase of Gas and Water Works,* 92–93.

51. Silverstone, *Purchase of Gas and Water Works,* 88–89; Local Government Board. *Urban Water Supply,* 86.

52. For example, in House of Commons debate, Dr. T. J. Macnamara, a Liberal, said, "Take Cardiff, Nottingham, Plymouth, Glasgow, Huddersfield, and Bradford, and Leicester—all these supply a wide area outside the municipal area . . . yet in [none] of these cases have the outside areas any representation [on] the Water Committees." *Hansard Parliamentary Debates,* 4th ser., vol. 103 (1902), col. 1330. In earlier years, the LCC had considered giving outside county councils seats in a specially constituted LCC water committee as part of the water solution, but after conferring with their counterparts, the outside councils decided that they were more interested in remaining independent from the LCC—and especially in not being obliged to contribute toward future costs of securing new sources.

53. Quoted in *Hansard Parliamentary Debates*, 4th ser., vol. 101 (1902), col. 1334.

54. Ibid., col. 1333.

55. Thomas Avery, *The Corporation of Birmingham and the Water Supply of the Town: A Statement Addressed to the Members of the Town Council* (Birmingham: The Journal, 1869), 14, 15.

56. Long, *Memories*, 135.

57. Councilor Dale, in Town Council debate reprinted in the *Bradford Observer*, 19 August 1852, 6.

58. *Hansard Parliamentary Debates*, 4th ser., vol. 103 (1902), cols. 1336, 1339–40.

59. Ibid., cols. 1361, 1365.

60. Ibid., col. 1354.

61. Sidney Webb, *The London Programme* (London: Swan Sonnenschein & Co., 1891), 36.

62. Minutes of Proceedings of a Deputation to the Local Government Board from the Water and Parliamentary Committees of the London County Council, Thursday, 13 December 1900, 4, LCC/MIN/12,276.

63. Ibid., 3.

64. Ibid., 5.

65. W. H. Dickinson, *The London Water Question*, Progressive Leaflet no. 2 (London: Progressive Election Committee, 1901), 18.

66. T. McKinnon Wood, *The L.C.C.: Three Years' Progressive Work* (London: Progressive Election Committee, 1901), 11.

67. *London and its water supply: Why has not London freed itself from the clutches of its Water Companies?*, Progressive Leaflet no. 6 (London: Progressive Election Committee, 1901), 4.

68. Quoted in *Star*, 3 February 1902, 3. The interview was given the sensational title, "Water Tyranny."

69. Quoted in *Morning Leader*, 18 February 1902, 5. Progressives contended that the bill was designed to undermine the LCC many times through many channels. "The main feature of the bill introduced by Mr. Long on behalf of the Government . . . is that it shows once more how the Government is still pervaded with the fixed idea that its duty is to suppress, if not destroy, the County Council as a municipal authority in London," wrote Dickinson in the *Morning Leader*, 1 February 1902, 4. "The Bill . . . prevents the County Council from being the water authority, and for purely political and in some cases commercial reasons," argued Burns, in "Commons debate," *Hansard Parliamentary Debates*, 4th ser., vol. 101 (30 January 1902), col. 1401.

70. "Commons debate," *Hansard Parliamentary Debates*, 4th ser., vol. 101 (30 January 1902), col. 1423. "It is necessary to have a large Board in which the

County Council will be out-voted. . . . Except that this Bill is framed mainly out of prejudice to the County Council, I cannot see any reason at all why this Board should have been created so large, unwieldy, and unrepresentative," added John Burns in debate just before the bill was sent to committee. *Hansard Parliamentary Debates*, 4th ser., vol. 104 (3 March 1902), col. 228.

71. London County Council Water Committee, Minutes of Proceedings, 5 February 1902, LCC/MIN/12,266, London Metropolitan Archives.

72. *Morning Leader*, 1 February 1902, 4. In a comprehensive pamphlet, W. H. Dickinson argued that "the new Trust will be more irresponsible and less easily affected by popular desire than the directors of the existing Water Companies, who at any rate are . . . under financial obligations to their shareholders." *The London Water Bill, 1902: A Protest on Behalf of Popular Government* (London: London Liberal Federation, 1902), 12. Dickinson was then chair of the London Liberal Federation.

73. "Commons debate, House of Commons, Thursday, 27th February 1902, London Water Bill, Second Reading," 49.

74. See, for examples, *Daily Chronicle*, 3 February 1902, 4; and two instances in the *Morning Leader*, 18 February 1902, 5.

75. *Hansard Parliamentary Debates*, 4th ser., vol. 104 (1902), col. 207; "Commons debate, House of Commons, Thursday, 27th February 1902, London Water Bill, Second Reading," 134.

76. London County Council, Debate on the Report of the Water Committee on the London Water Bill, 1902, 11 February 1902, transcript for shorthand notes, 15–16, F/DCK/14/3, London Metropolitan Archives.

77. "Commons debate, House of Commons, Thursday, 27th February 1902, London Water Bill, Second Reading," 138. See also *Hansard Parliamentary Debates*, 4th ser., vol. 104 (1902), col. 216.

78. London County Council Water Committee, Minutes of Proceedings, 22 July 1902, LCC/MIN/12,266.

79. W. H. Dickinson, *The London Water Bill, Letter to the "Daily News" by Mr. W. H. Dickinson, being a Protest against introducing a Foreign System of Government into English Local Administration* (London: National Press Agency, 1902), 1.

80. *Hansard Parliamentary Debates*, 4th ser., vol. 101 (30 January 1902), col. 1382.

81. "House of Commons. London Water Bill. Debates on the motion for leave to introduce the Bill and on the second reading," 36. See also *Hansard Parliamentary Debates*, 4th ser., vol. 104 (1902), 242.

82. Proceedings of a meeting of the East St. Pancras Conservative Council, printed in *St. Pancras Guardian*, 28 February 1902, 7.

83. *City Press*, 15 February 1902, 4; *East London Observer*, 15 February 1902, 5.

84. *Hansard Parliamentary Debates*, 4th ser., vol. 104 (1902), cols. 257–65.

85. House of Commons, *Sessional Papers,* 1902, vol. 4, Minutes of Evidence of the Joint Select Committee, Questions 3589, 3649, 3737, 3775.

86. On this episode, see G. L. Gomme [LCC clerk] and Harry Haward [LCC comptroller], *London Water Supply: Report on the action of the Council with regard to the water supply of London with a special report by the Comptroller of the Council on the financial aspects of the purchase* (London: London County Council, 1905), 26; Mukhopadhyay, *Politics of Water Supply,* 148.

87. On this episode, see Mukhopadhyay, *Politics of Water Supply,* 146–48.

88. Local Government Board paper 130272, MH 29/47, National Archives, quoted in Mukhopadhyay, *Politics of Water Supply,* 149.

89. Campbell-Bannerman, speaking in House of Commons debate, quoted in Mukhopadhyay, *Politics of Water Supply,* 149.

90. Long, *Memories,* 137.

91. Henry Jephson, *The Sanitary Evolution of London* (London: T. Fisher Unwin, 1907). A year after the LCC's defeat was sealed, Bernard Shaw wrote a bit bitterly about the hostile Parliament and warned against the spread of the "plutocratic Collectivism of the Trusts" that, apparently, the Metropolitan Water Board represented. George Bernard Shaw, *The Common Sense of Municipal Trading* (Westminster: Constable, 1904), 58.

92. London County Council, Minutes of Proceedings, 1903, 17 February 1903 (London: LCC, 1903), 213.

93. Metropolitan Water Board, "Transcript of the shorthand notes of the first, second, and third meetings of the Board," item 221, Sanitary Reform of London Collection, Special Collections, Stanford University Library, Stanford, CA. In LCC debate, a council member named Dr. Napier "expressed regret that so many distinguished members of the Council were going on to this Water Board, against the strong feelings of some of them, who thought it would be exceedingly difficult for such a body to succeed in its objects. At any rate, whether it was a success or failure, it could never in the future be said that the Council had not accepted the position laid down by Parliament and done its best to make it a success." *Times,* 18 February 1903, 4.

94. Metropolitan Water Board, "Transcript of the shorthand notes of the first, second, and third meetings," 34.

95. Ibid., 25.

96. Ibid., 44, 62.

97. In practice, however, the first eight chairs refused their salary and accepted only an expense allowance. London Water Board, *London's Water Supply, 1903–1953* (London: Staples Press, 1953), 17.

98. Harry Haward, *The County Council from Within: Forty Years' Official Recollections* (London: Chapman & Hall, 1932), 341. Haward was the LCC's comptroller and participated in the arbitration.

99. See Gomme and Haward, *London Water Supply.*

100. Haward, *County Council from Within,* 344.

101. Mukhopadhyay, *Politics of Water Supply,* 155.

102. Haward, *County Council from Within,* 344.

103. On the expansion of works in the 1890s, see H. W. Dickinson, "Water Supply of Greater London, Part 14," *Engineer* 186 (October 1948): 354; Gomme and Haward, *London Water Supply,* 36; Mukhopadhyay, *Politics of Water Supply,* 158; Henry Jephson, *The Making of Modern London: Progress and Reaction* (London: Bowers Brothers, 1910), 51.

104. Dickinson, *London Water Bill, 1902,* 15.

105. Malcolm Falkus, "The Development of Municipal Trading in the Nineteenth Century," *Business History* 19, no. 2 (1977): 146.

106. John Hassan, *A History of Water in Modern England and Wales* (Manchester: Manchester University Press, 1998), 19–20; Falkus, "Development of Municipal Trading," 152.

107. John James Harwood, *History and Description of the Thirlmere Water Scheme* (Manchester: Henry Blacklock & Co., 1895), 106–7. See also Robert Farquhar, *Objections to the Thirlmere Scheme* (Ambleside: W. Porter and Sons, 1879); and Harriet Ritvo, *The Dawn of Green: Manchester, Thirlmere, and Modern Environmentalism* (Chicago: University of Chicago Press, 2009).

108. T. W. Woodhead, *History of the Huddersfield Water Supplies* (Huddersfield: Wheatley, Dyson & Son, 1939), 58. Thomas Hawksley consulted on the project, but it was not his scheme.

109. Stephen Inwood, *A History of London* (New York: Carroll & Graf, 1998), 412.

110. Long, *Memories,* 99.

Conclusion

1. Municipalization would probably have been affordable, given the council's ability to secure loans at a 2.5 percent interest rate—less than the Metropolitan Water Board's interest rate—and given the fact that the water companies' existing rates secured them a profit. Taken as a whole, municipal water enterprises in 1907 in England and Wales yielded a 5 percent profit on cities' expenditures, and most of that profit was invested in lowering rates paid by consumers. Douglas Knoop, *Principles and Methods of Municipal Trading* (London: Macmillan, 1912), 311.

2. Sidney Buxton, *Hansard Parliamentary Debates,* 4th ser., vol. 103 (27 February 1902), col. 1322.

3. See, for examples, *Times,* 1 July 1905, 11; 8 July 1905, 5; 25 May 1907, 5.

4. Provincial town councils experienced the same challenges. On Liverpool's series of projects undertaken to meet constantly growing demand, see Derek Fraser, *Power and Authority in the Victorian City* (New York: St. Martin's Press, 1979), 44–46.

5. Anne Hardy, "Water and the Search for Public Health in London in the Eighteenth and Nineteenth Centuries," *Medical History* 28, no. 3 (1984): 282.

6. The model of an ecosystem in part derives from Jared Orsi, *Hazardous Metropolis: Flooding and Urban Ecology in Los Angeles* (Berkeley and Los Angeles: University of California Press, 2004), esp. 1–11.

7. On the changes in climate, see H. J. Fowler, C. G. Kilsby, and P. E. O'Connell, "Modeling the Impacts of Climatic Change and Variability on the Reliability, Resilience and Vulnerability of a Water Resource System," *Water Resources Research* 39, no. 8 (2003): 1222–33.

8. Fred Pearce, "Water Anger Everywhere but No One Stops to Think," *Independent,* 27 August 1995; Joanna Bale, "Wasted Water Blamed for Hose Ban," *Times,* 3 April 2006.

9. Fiona MacRae, "Britain's Biggest Drips," *Daily Mail,* 16 March 2006; "News Analysis: Drought Britain," *Independent,* 19 March 2006; "Outrage as Thames Water Escapes Fine," *Daily Mail,* 4 July 2006.

10. On anger over the expansion of reservoirs in the Thames valley, see Andrew French, "Credit Crunch Delays Thames Water Plan," *Oxford Mail,* 3 March 2009. On calls for the Thames Water Company to find a new source of water in Wales, see C. S. McCulloch, "The Water Resources Board: England and Wales' Venture into National Water Resources Planning, 1964–73," *Water Alternatives* 2, no. 3 (2009): 473. On calls for a national water grid for Britain in lieu of private regional utilities, see Sean Poulter, "Only a National Water Grid Will Spare the South from Drought, Warn Engineers," *Daily Mail,* 27 June 2006; and personal communication with members of GARD (Group Against Reservoir Development).

Index

Abbey Mills lift station, 175n74
Albert Embankment, 47
alcohol, x, 7, 40, 172n37, 172n39
Alison, William Pulteney, 67
Anglican Church, 4, 43
Ashton-under-Lyne, 133
Asquith, H. H., 140
Avebury, Lord, 123. *See also* Lubbock, Sir John

Balfour Commission, 61, 90–91, 92, 95, 112
Balfour, Lord, of Burleigh, 91, 112, 141
Balian, George, 95
Bank of England, 46
Barry, Sir John Wolfe, 199n42
Bateman, John Frederic La Trobe- (1810–89), 29, 37, 85, 166n124, 176n93; principles of water-works design, 23–24, 62, 176n93; gravitation schemes, 20–22, 26, 133, 166n122; Welsh scheme and, 49–54, 83, 100–101, 105
Bazalgette, Joseph, 46, 48, 51–54, 59, 149

Beal, James (Westminster vestry-man), 54–60, 152
Belfast, 2
Bentham, Jeremy, 6–7,
Benthamite utilitarianism, 75,
Besant, Annie, 70, 75, 181n6
Binnie, Alexander (engineer), 118, 190n95; testimony to LCC, 110–11, 113–14; Welsh scheme and, 83, 100–101, 104–7
Birmingham, 86, 110, 134–35; compe-tition for Welsh water, 101–4; municipalization in, 89, 108, 125, 145, 169n165, 191n106, 199n39; re-formers of, 83–84, 87, 183n58. *See also* Chamberlain, Joseph; Elan Valley water development scheme
Blackwall pedestrian tunnel, 195n164
Bland, Hubert, 76
bleachworks, 3
Board of Health (created 1848), 27, 42, 160n36, 162n57. *See also* Chad-wick, Edwin
Board of Trade, 44, 91
Boer War, 130
Boulnois, Edmund, 136,